F. L Stewart

Sorghum and its Products

F. L Stewart

Sorghum and its Products

ISBN/EAN: 9783337292713

Printed in Europe, USA, Canada, Australia, Japan

Cover: Foto ©Suzi / pixelio.de

More available books at **www.hansebooks.com**

EVAPORATING RANGE.—Fig. 1.

SORGHUM AND ITS PRODUCTS.

AN ACCOUNT OF RECENT INVESTIGATIONS

CONCERNING THE

VALUE OF SORGHUM IN SUGAR PRODUCTION,

TOGETHER WITH A DESCRIPTION OF

A New Method of Making Sugar and Refined Syrup

FROM THIS PLANT.

ADAPTED TO COMMON USE.

BY

F. L. STEWART.

PHILADELPHIA:
J. B. LIPPINCOTT & CO.
1867.

TO

JOHN FRASER,

PRESIDENT OF THE AGRICULTURAL COLLEGE OF PENNSYLVANIA,

This Work

IS RESPECTFULLY INSCRIBED

BY HIS FRIEND,

THE AUTHOR.

PREFACE.

RECENT improvements in the art of making sugar and syrup from Sorghum Cane have created a necessity for the publication of this volume. It is written for the benefit chiefly of Farmers and Planters of the United States, who have fostered this new branch of industry from the outset, and in whose hands it is destined to assume a new importance. The time is past when this pursuit is to be regarded only as an experiment, or as an ephemeral enterprise, persisted in merely because of the paralyzation of industry at the South.

Sugar making is an Art, and complete success is not attainable at a bound, but the attentive reader of these pages will find that it is an art easily learned. The pathway, by which I reached my object in these researches, was beset with obstructions which it was necessary to overcome successively as they presented themselves—wherein much time was spent, but not unprofitably, as the result, I trust, will show. Of the magnitude of these difficulties few have any just apprehension, who have not, in a thorough manner, addressed themselves to the task of discovering what is their real nature.

One fact, however, rewards the investigation at the start, and was to me a sufficient encouragement; namely, that the best varieties of this *cane contain as large a proportion of crystallizable sugar as does ordinarily the sugar cane in Louisiana.* This

1*   (v)

truth has not generally been recognized, but the sooner it becomes known and appreciated the better.

In common with all who would obtain accurate information respecting sugar production at the South, I acknowledge my indebtedness to the works of Evans and Dutrone, and to the more recent and very valuable researches of Professor R. S. McCulloh.

With the confident expectation that the processes herein recommended will be successful in other hands, as they have been in my own, the work is submitted to the public.

F. L. STEWART.

April, 1867.

CONTENTS.

CHAPTER I.

SUGAR AND SUGAR PLANTS.

CHAPTER II.

METHOD OF CULTIVATION.

CHAPTER III.

METHOD OF CULTIVATION (CONTINUED).

CHAPTER IV.

METHOD OF CULTIVATION (CONTINUED)—EARLY PLANTING.

CHAPTER V.

METHOD OF CULTIVATION (CONTINUED)—RIDGING.

CHAPTER VI.

MANURES.

CHAPTER VII.

MANURES (CONTINUED).

CHAPTER XII.

HARVESTING THE CANE.

CHAPTER XIII.

STORING THE CANE.

CHAPTER XIV.

PROCESS OF MANUFACTURE.

CHAPTER XV.

PROCESS OF MANUFACTURE (CONTINUED).

CHAPTER XVI.

PROCESS OF MANUFACTURE (CONTINUED).

CHAPTER XVII.

PROCESS OF MANUFACTURE (CONTINUED).

CHAPTER XVIII.

PROCESS OF MANUFACTURE (CONTINUED).

CHAPTER XIX.

SUGAR MILLS.

2

CHAPTER XXIX.

NATURAL AFFINITIES OF THESE CANES—INFERENCES DERIVED THEREFROM.

CHAPTER XXX.

OTHER PRODUCTS.—SYNOPSIS OF PROCESS OF MANUFACTURE.

SORGHUM AND ITS PRODUCTS.

CHAPTER I.

SUGAR AND SUGAR PLANTS.

Cane Sugar—Its Early History—The Period of its Introduction into Europe—Comparatively unknown to the Ancients—Pliny's Observations—Importance of the Sugar Industry at the Present Time—Sources from which Sugar is now derived—The Tropical Cane—Its Climatology—Sugar Production in the United States—Insufficiency of all our Native Sources of Supply—The Cane in Louisiana—The Sugar Maple—The Beet—Grape or Starch Sugar—Its Low Rank—Inapplicable to the most Important Purposes for which Cane Sugar is used—Sorghum Saccharatum, the Northern Cane—Sketch of its History—Its Relations to Soil and Climate—Its True Character and Value.

THE introduction of new alimentary substances and skill in the art of producing them in the forms best adapted to human use, are eminently characteristic of modern times. There seems to be truth in the opinion that the vigor and activity of intellect and superior power of physical endurance possessed by the Caucasian race are due, in some sort, to the use of certain articles of diet unknown or unappropriated in the earlier ages. It is at least a notable coincidence that the period when Anglo-Saxon power and civilization began to be acknowledged in the world, the golden era of literature and philosophy in Europe, the age of Shakspeare and Bacon, was marked by a dietetic

(15)

revolution, in the progress of which, tea, coffee, and sugar came into general use.

Sugar was almost unknown to the Greeks and the Romans. By our ancestors of the north of Europe it was occasionally used as an agreeable condiment only; its place, both in ancient times and during the middle ages, being supplied in part by honey. So ignorant were the ancients of any system of manufacturing it, that they supposed it to exude naturally from canes, like gum. Even Pliny in his Natural History alludes to it but very briefly in his enumeration of the rare productions of the East, observing that a substance called "saccharen" was obtained from certain reeds in India; that it was of a white color, was sweet like honey, crackled like salt between the teeth, and was found in lumps of the size of a hazel-nut. This could have been nothing else than the rock candy, or crystallized white sugar, of the kind yet made in Cochin China. The Greek physicians bought it occasionally from the Arabian merchants, and used it as a medicine. From India, where it was early cultivated, the sugar cane was carried by the Arabians to Mesopotamia, a country celebrated for its sugar at the period of the Crusades, whence it was successively introduced into Syria, Egypt, Sicily, and Spain.

Sugar has become one of the most indispensable products of modern industry, and any means by which its production may be facilitated, or increased, is now of importance to the whole human race. In our own country, the consumption of sugar is much greater in proportion to our population than anywhere else on the globe, and our interest in the supply of this product is still further enhanced by the consideration that it not only falls far short of the demand for it, but also that, prospectively, the disproportion between the supply through the old channels and the demand is likely to be much greater.

Sugar plays an important part in the economy of nature, animal and vegetable; and although our supply of it has been mainly derived from the tropical sugar cane (*Saccharum officinale*), the natural sources of it are as numerous and as widely distributed over the earth's surface as the bread plants; no region being found destitute of some species of sugar plants, or uncongenial to them, in which any of the cereals can be successfully grown. Sir John Richardson found the ash-leaved maple, or box-elder, at the 54th parallel, affording most of the sugar made in Rupert's Land, which may be regarded as the extreme northern limit to the successful growth of wheat and barley in the valley of the Saskatchewan. In lower latitudes, commencing at the parallel of 50°, the maples, all sugar-producing trees, and some varieties of sorghum and sugar beet lately introduced, are found coincident in range with the maize and the less hardy grains, until, at last, within the rice-belt of the South we encounter the Indian cane, and the sugar palms.

Cane sugar is also contained in a great many other plants from which it has not been extracted, as yet, in sufficient quantity to render its manufacture profitable; among these may be mentioned various grasses, Indian-corn before it has ripened its grain, bulbous roots, such as the turnip, parsnep, and carrot, some fruits, as the pumpkin, melon, banana, etc., the fruit of the European chestnut, the nectary of the flower of Rhododendron ponticum, and, at certain periods, in the sap of some trees, such as the walnut-tree of the Caucasus, the American white walnut, hickory, birch, etc. The sugar obtained from these sources is identical in composition and sensible properties with that contained in the juice of the tropical cane, and it is the only kind of sugar applicable to ordinary use.

Other saccharine substances or sugars of a different and

inferior kind are common, and especially that which constitutes the sweet principle in most of our cultivated fruits,
of which the sugar of the grape may be regarded as the
type.   This sugar has been produced in the solid form to
a small extent directly from the inspissated juice of the
grape in Syria, Egypt, and some parts of France.   It is
identical with the finely granulated coating found upon
the surface of raisins, and with the mealy sediment gradually formed upon the bottom of vessels containing honey.
Without alluding to the chemical properties and relations
of this substance, which are more appropriately reserved
for discussion in another part of this volume, it is enough
to say that grape sugar is entirely different from true cane
sugar, and can never be substituted for it in domestic use.
It is soft, mealy, and liable to become damp and to fer-
ment.   It curdles milk, and is inappropriate for sweetening
coffee and tea, and for most culinary purposes.   In saccharine richness it is far inferior to cane sugar, two pounds
of the latter being equal in sweetening properties to five
of the former.   Of the same kind is the sugar artificially
prepared from starch.   It is manufactured from the starch
of the potato in Europe; and recently an attempt was
made in New York City to produce it from corn starch
under the name of corn sugar.*

The exorbitant prices now demanded for sugar, resulting
chiefly from the partial failure of our Southern sugar crop
during a period of several years prior to the recent rebellion, and the utter prostration of the business of production
since, together with a vastly increased demand for sugars,
is an evidence of our humiliating dependence upon foreign
countries with which we formerly competed with success.

---

* Corn sugar is the name appropriately given to sugar produced from
the juice of maize or Indian-corn, and it is identical with cane sugar.   The
name as applied to starch sugar is liable to mislead.

During the year 1855, when the degeneracy of the cane in Louisiana began to become apparent, 822,000,000 pounds of cane sugar were consumed in the United States, of which 440,000,000 were of foreign importation, and 382,000,000 of American growth; more than three-fourths of the latter amount were supplied by Louisiana. Owing to local causes, the annual production in Louisiana had been rapidly diminishing since 1853. During that year the crop was estimated at 400,000,000 pounds. In 1854 it was 346,500,000 pounds. In 1855 it was 231,426,000 pounds, and subsequently the yearly average was much less.

Nature has set barriers to the geographical range of the southern sugar cane beyond which it cannot be grown with success. The extreme limits of its distribution north and south, as far as determined by general climatic influences, appear to be the parallels of 30° on each side of the equator. The soil and local peculiarities, however, of the region comprised within these boundaries are not always favorable. The regions from which the markets of the world are supplied are neither numerous nor extensive. They are, chiefly, the West and East Indies, British India, and the Island of Mauritius. Within the United States, we cannot hope to extend the limits of growth of this plant beyond the narrow belt of territory along the Gulf shore, where it has been already planted, and, even there, causes are at work inducing disease and deterioration of the cane, which, unless they can be checked, will entirely prevent the resuscitation of the business of sugar production from the tropical cane.

We turn in vain to other native sources of supply. The sugar maple of our Northern forests is utterly inadequate: 40,000,000 pounds of maple sugar are annually produced, but that quantity is but a small fraction of the amount required to meet the increased demand. The manufacture of sugar from the beet, although successful in Europe, has not been fully tested in this country; but on account of the great expensiveness of the machinery, the high degree of skill necessary, the radical changes which it would inaugurate in the means and methods of agricultural labor already existing, the inferior quality of the secondary products (no molasses being made from the beet which is fit for human use), and especially the great length of time necessary to establish it upon a solid basis, place it beyond our reach, and render it unsuited to meet our wants.

It is just at this juncture that our attention has been called to a plant which seems as adequate to supply us in future with sugar, as, in the few years that have elapsed since its introduction into our midst, it has proved itself to be capable of providing for half the tables in the land an abundance of rich and palatable syrup. This plant is the sorghum. It is called in botanical terms *Sorghum saccharatum;* all the different kinds being now recognized as varieties of one species.

The Chinese sorghum was imported into France from the north of China about the year 1851. Through the agency of the Patent Office it was obtained from France in 1854, and during the spring of the following year the seeds were distributed to different parts of the Union. The success which attended the first efforts to make a palatable article of syrup from the juices of this plant awakened attention, and in 1857, when public curiosity was at its height, Mr. Leonard Wray arrived in this country, bringing with him the seeds of fifteen varieties of South African sorghum, or imphee. These he first found growing in the country of the Zulu Caffres, near Cape Natal, in the year 1851.

Subsequent experiments made upon these canes, grown by him in South Africa, were rewarded with success in the production of sugar. The early history of these varieties is buried in profound obscurity, although there is reason to believe that some of the saccharine sorghums are alluded to in the writings of the old authors. In reference to this, Mr. Wray observes :

"If we look back from our own times to very remote ages, and search for any very authentic records of the imphee, or *Holcus saccharatus*, among the writings of ancient authors, we must confess the unsatisfactory results of our inquiry ; for the notices of the 'sweet reed' contained in their writings have long ago been seized by Porter and other authors, and appropriated by them as forming part and parcel of the history of the *sugar cane*.

"But if we examine somewhat minutely into the matter, we shall find abundant reason for believing that the *Holcus saccharatus* was frequently alluded to, instead of the sugar cane, more especially by the Roman writers ; thus Lucien (Book iii. page 237) has the line—

"'Quique bibunt tenera dulces ab arundine succos,'*

which can scarcely be supposed to apply to the *large, coarse, hard stalk* of the sugar cane.

"Besides this, we all know that the Romans had a very excellent general knowledge of the products of Ethiopia, in which varieties of the *Holcus saccharatus* are to be found ; and they, no doubt, knew that the natives ate, or rather chewed its stalks, for the 'sweet juices' contained in them.

"The native traders, who took a coarse kind of *goor* or *jaggery* to Muciris and Ormus, always said that they obtained it from a 'reed ;' and I have no doubt that they did really obtain it from this reed-like plant."

---

* "And those who drink sweet juices from the tender reed."

Some variety of imphee had been introduced into Europe as early as a century ago, but it was evidently inferior to the imphee cane now cultivated in this country. Its seeds were described to be of a clear brown color. In 1766 it was experimented upon for the extraction of sugar by one Pietro Arduino, at Florence, in Italy,* but with what success we are not informed.

The following extract from Reese's Family Encyclopedia (p. 727), no doubt refers to the same or a closely allied variety :

"A large grass (*Holcus cafer*), brought from the south of Africa, has begun to be cultivated in some parts of Italy, Bavaria, and Hungary for sugar, and what is made from it equals cane sugar."

M. Vilmorin also says, that in a collection of plants sent to the Museum of Natural History, at Paris, in 1840, by M. d'Abadie, there were thirty kinds of sorghum, among which he particularly recognized several plants having stems of a saccharine flavor.†

The almost total seclusion from foreign intercourse of that portion of the Chinese Empire in which sorghum is grown, has left us well-nigh destitute of information respecting our most important variety, if we except some scattered waifs, which are generally too meager to serve any useful purpose.

The Rev. Justus Doolittle, whose book on the "Social Life of the Chinese"‡ has been lately issued, says :

"The so-called Chinese sugar cane or sorghum is grown very extensively in Northern China, and is known among foreigners as a kind of millet—the Barbadoes millet. The Chinese name for it is Kauliang.   *   *   *   The

---

* Mason's Circular, Dec. 10, 1856.

† Mason's Circular, Dec. 10, 1856.   ‡ Harper & Bros., 1866, p. 43.

Chinese do not express the juice from the stalks for the purpose of manufacturing molasses or sugar, and they manifest surprise when informed such a use is made of it in the United States. They make a coarse kind of bread from the flower of the seeds of the kauliang, eaten principally by the poorer classes. The best kind of Chinese whisky, often called Chinese wine, is distilled from the seeds. The stalks are used for food or lathing in the partitions of houses, for slight and temporary fences, etc. During a few years past, many inquiries have been made in regard to the manner in which the Chinese manufacture sugar and molasses out of the sorghum, but such information is vainly sought of them."

When it is remembered that in those parts of Southern China in which the tropical cane is grown, the natives do not generally use that plant for making sugar; but adopt the more primitive practice of peeling and chewing the stems, just as the Zulu-Caffres use the imphee in South Africa, we need not be surprised to hear that in the northern provinces of the empire nothing is known of any method of making sugar from sorghum.

Another author, however, asserts that in the country near Shanghai, "sugar is made from the cane which is now well known in the United States as the 'Chinese sugar cane,' and is extensively used in making confections, sweetmeats, and preserves, of which the ginger put up at Canton in small blue jars is most familiar to us."[*] It is also said that in 1853 a number of sacks of sorghum sugar, of seventy-five pounds weight each, were imported into California from China.[†]

---

[*] "Five Years in China," by Charles Taylor, M.D. (former missionary to China). New York, 1860, p. 131.

[†] E. F. Newberry, Essay on Sorghum. Valley Farmer, 1863.

The Abbé Huc, a traveler in Thibet and Tartary, says the brandies of the North are made principally from a large millet (*Holcus sorghum*). It is probable that when we gain access to all the knowledge possessed by the Chinese as to their mode of cultivating this plant, and of crystallizing its juice, we shall not value the information very highly, and it is of more importance to know what are the fruits of careful study of the nature of this cane as it is now found in our own country, and of chemical research as to its value in the production of sugar.

CHAPTER II.

METHOD OF CULTIVATION.

The Method of Cultivation adapted to the Nature and Requirements of the Plant—Resemblance to Indian-corn not so close
as has been supposed—Cereal and Saccharine Plants—Importance of adherence to System—Some Considerations of Primary
Importance—The Selection of the Seed—Varieties of Cane—
Chinese and South African—Characteristics of Good Seed—How
Germination may be hastened—Physiology of the Seed—Planting and Cultivation—Outline of a System defined.

THE resemblance of sorghum to Indian-corn has been so
far exaggerated as to lead some to the inference that the
same mode of cultivation is equally well adapted to both.
This opinion cannot be sustained, for it practically ignores
certain characteristics and requirements in respect to which
they widely differ. Experience has in some measure cor-
rected this error; but the easily learned rule to "plant and
cultivate the same as Indian-corn," is yet too much in
vogue, and strict adherence to it has, more than anything
else, hindered improvement in the cultivation of the plant,
and in the development of a more rational system especially adapted to its nature.

When it is considered that the relationship naturally
existing between the cane and common corn is not
more intimate than that expressed in the fact of the possession by each of certain general characters constituting
them members of the same great family (the grasses); that
they are both specifically and generically distinct; that
they are grown for widely different purposes, and that

some of the conditions essential to the perfection of the one
as a grain-bearing plant, are inimical to the special function
of the other as a sugar producer; that there are other constitutional peculiarities in which they differ, such as the
adaptation of the cane to a lower temperature during the
early period of its growth than corn, its greater capacity
to endure drought after its growth has become well established, chiefly on account of its deriving its nourishment
at that time from a lower stratum of the soil, its ability to
endure more frost, both early and late, and its greater sen-
sitiveness to certain peculiarities of climate and soil,—it
is obvious that conformity in the mode of culture to these
distinctive properties at least is essential.

Strict adherence to a well-defined system is imperative
throughout the whole progress of sugar production, and
at no time more so than during that period when the subtle
chemistry of soil and air is at work within the cells of the
growing plant, elaborating and distributing the materials
upon which the skill and labor of the manufacturer are
afterward to be expended.   It should never be forgotten
that neglect at any period will inevitably be followed by a
forfeiture of the best results.

At the outset, great caution should be observed in the
selection of the seed for planting, so that not only the
richest and purest varieties of the cane, but those also best
adapted to the climate of each district, may be secured.
To a want of discrimination in this respect is to be attrib-
uted much of the ill success which has occasionally attended the experiment with the sorghum in this country.

The different varieties which were distributed over the
country a few years ago have exhibited very different degrees of acclimatization. Sufficient time has not yet elapsed
to justify a definition of the climatic range of each of these.
When fully naturalized, some of them will have acquired

more vigor (a result already noticeable), and will be
enabled to withstand a more rigorous climate than at
present; and new varieties will, no doubt, be produced
which may be grown with profit in districts from which all
are now excluded.   At present more flexibility of constitution seems to be possessed by the Chinese cane than the
imphees.   Some of the latter are analogous to those
northern varieties of the maize of which the King Philip
corn is a representative.   They come to maturity within
a comparatively brief period, and, although of less gen-
erous growth and proportionally less productive, are
not inferior to the best southern varieties in the quality of
their products.   They are adapted to the briefer summer
of more northern latitudes; and the preference which some
planters in the Northwestern States have accorded to them
is to be ascribed to the fact that they ripen there and mature their juices, while other varieties do not.   The Chinese
cane, however, and some of the imphees likewise, plants
of a more luxuriant and permanently sub-tropical type,
bear a strong analogy to those stately races of the maize
which attain to such unequaled perfection in the middle
belt of the United States.   Within that region (elsewhere
more fully defined)* are now grown varieties of cane which,
if unhybridized, we may expect to attain to the highest
development of which the species is capable.   In more
northern localities, early maturing imphees, such as the
Ne-a-za-na, Oom-see-a-na (misnamed Otaheitan), and
E-en-gha are to be preferred.

It is worthy of note that the qualities of the future plant
will depend in a great degree not only upon the purity,
but also upon the proper development of the seed.   The
largest, heaviest, and best formed seeds uniformly produce

---

* Chapter IX.

the strongest and healthiest plants. Such seeds contain
the most starch, a substance which undergoes a peculiar
transformation during germination, and which supplies the
young plant with food until it has sent out its rootlets and
leaves. The earliest ripened and heaviest seeds are always
to be found on the summit or upper half of the panicle;
and, from such, a selection should in all cases be made,
rejecting those from the lower half which are often not
ripened fully, and always imperfectly filled, and which produce stunted and ill-developed plants, capable of transmit-
ting only their own inferior qualities.

Before planting time, the germinating power of all seeds
not previously well known, should be tested by placing
them on moistened blotting-paper or cotton, in a warm
place, excluded from light, till they sprout. This method
is better than planting in soil, as the progress of growth
may be readily noticed, and as different depths of soil variously influence the result.

Growth may be hastened by steeping the seed and in
other ways; but when a system of early planting is practiced, the use of such means is neither necessary nor desirable. Yet, in all cases where-delay is unavoidable, the
most successful mode of preparation is to put the seed in
a sieve or coarse sack, and pour hot water upon it. It
may be steeped in the water for twelve hours or more, and
then rolled in plaster or ashes and planted.

Various chemical agents have been employed with the
object of accelerating germination, by dissolving them in
the water in which the seed is soaked. The action of
these substances, however, seems to be most energetic, not
during germination, but afterward, when the earth in
which the seed is planted is irrigated with their solutions.
In the seeds of all plants, the first movement of the
embryo is determined by its exposure to a certain degree

of heat, the requisite moisture also being supplied. The first food which the germ receives is contained in the seed itself, and there is no evidence that water holding in solution any of the salts that have been used for this purpose hastens germination, or facilitates the transformation of the starch in the seed in any greater degree than water alone. These, as well as other liquid manures, may be applied, however, with good effect when the roots and leaves have been well formed, and the supply of nutriment in the seed has been exhausted. Steeping in hot water is probably the easiest and most certain method of applying heat, so as to induce quick germination; but if too long continued, it will either impair greatly the native hardihood of the young plant, or completely destroy its vitality; therefore it should be practiced only when early planting is impossible.

3*

CHAPTER III.

METHOD OF CULTIVATION (CONTINUED).

Inequality of Developement of the Stem and Roots of the Annual
Grasses during the early Period of Growth—This Peculiarity
must be recognized in any Rational System of Planting and
Culture—All Cultivated Plants divided into two Great Classes,
according to their mode of Development—Relations to Climate—
The Oat Plant—Ahrend's Experiments—In what Respects the
growth of Sorghum is Analogous—The Characteristics of the
Plant at different Periods of Growth exhibited—Physiology of
the Leaf, Stem, and Root—Treatment which the Plant should
receive at each Period.

In the planting and culture I would recommend:

1. Fall plowing, and deep and thorough tillage.

2. Very early planting.

3. Ridge planting as distinguished from planting in a
furrow or drill on a flat surface.

The above, together with other important but subordinate particulars, form a series of operations inseparably connected, the whole constituting a system of culture which
derives its chief value from a close conformity to the real
nature and requirements of this plant. Let us now see
what these are.

Annual plants as to their mode of development may be
divided into two great classes, viz.: 1st. Those like the pea
and the tomato, in which the growth of the stem and leaves
is directly proportioned to that of the root during every
stage of their existence. Like perennials, their development is equal throughout.

(30)

2d. Those, like most of the annual grasses, which exhibit during certain periods no such equality of growth in the different parts. Instead of adding uniformly to every part of their structure the organized matter as fast as it is formed, they are busy at certain periods of their existence in accumulating it in special parts, from whence it is distributed with great rapidity at particular periods to other parts toward which the vital energies of the plant seem suddenly to be diverted. In this respect, such plants bear a strong analogy to biennials, which, during the first year of their growth, accumulate in their roots the materials from which the luxuriant growth of the second season is rapidly evolved. Of this class sorghum is a conspicuous representative, as are also, to a certain degree, all our cereal plants.

The relation which this peculiar mode of development bears to the climate of the different parts of the year is one of those beautiful examples of adaptation which meet the eye of the observer in every part of the kingdom of nature, and a proper knowledge of it is of primary importance in all our attempts to mark out a suitable plan of cultivation. Ahrend's experiments with the oat-plant, and those of Anderson with the turnip, exhibit the strong analogy which an annual of the class last described, in the early period of its growth, bears to a biennial plant, during the first year of its existence.* In both, before the close of a certain period after germination, the leaves of the young plant "lose the power of applying to their further growth the food which they had absorbed, and which now, transformed into organizable matter, was deposited in the roots. The same nutritive particles which went to form leaves,

---

* Liebig. Nat. Laws of Husbandry.

so long as the mass of foliage kept on increasing, now became constituent portions of the root."*

The growth of sorghum is precisely similar. The apparently half-dormant period after the young leaves have attained their development is that during which "the migration of the constituents of the leaves and transformation into the constituents of the root" is taking place. That such a transfer really occurs at this period is evinced by the great bulk of root surface as compared with the foliage, the former increasing most rapidly and extending itself farthest during that stage when the leaves are "at a stand." At its close the tuft of young leaves has a height of only 5 or 6 inches, while the roots are extended widely with a length of 2 or 3 feet. This stationary period, as respects the early leaf growth of sorghum, comes to a close in about 60 days from the time of germination.

A second stage of growth immediately follows this, which is characterized by the most remarkable activity in that part of the plant which previously seemed almost dormant, and is produced by a correspondent change in the direction and intensity of vital action. There is now commenced a sudden and rapid transportation of the nutritive matter from the root to the stem, which begins to shoot up, and all the vital functions at the same time are wonderfully quickened. The measure of this activity, however, depends upon the character of the season. This stage is completed when the stalk of the cane has attained to nearly its full size. Three well-defined periods of development successively follow, embracing severally the flowering season, that in which the seed is formed, and that of ripening.

---

* Idem.

The relative length of each of these periods of growth in oats and sorghum (Chinese) is about as follows :

			OATS.	SORGHUM.
1st period.	From germination until before shooting............................		49 days.	60 days.
2d "	From shooting of the stem until it is full grown (in flower)....		12 "	30 "
3d "	Flowering..............		10 "	15 "
4th "	Formation of the seed..............		11 "	10 "
5th "	Ripening.............................		11 "	15 "
	Total...........................		93 "	130 "

A careful study of the phenomena of growth manifested by these plants at different stages of their existence will reveal the fact that they bear a definite relation to conditions of climate prevalent only during regularly recurring periods in the progress of the seasons.

Throughout the whole range of organic life, to plants as well as to animals, nature has imparted certain modes of existence and peculiarities of structure adapted to the physical influences in contact with which they were designed to live. These essential differences must in all cases be provided for by the planter. That method of culture, therefore, will be most nearly a true one by which these natural requirements of the plant are the most fully satisfied.

CHAPTER IV.

METHOD OF CULTIVATION (CONTINUED)—EARLY PLANTING.

Early Planting essentially Important, and why—Analogy to Wheat—"Volunteer Cane"—Conditions of Growth during the First Period, and Influence of a Low Temperature—This Period falls within the Limits of the Meteorological Spring, in the Latitude of Philadelphia—Disadvantages of Planting at a Later Period—Reasons why a System of Early Spring Planting should in all Cases be followed—How the Conditions necessary to Success by this System may be obtained—Fall Plowing and Deep Tillage almost indispensable—Collateral Advantages secured thereby.

WHAT mode of treatment, then, does the sorghum demand during the several periods of growth already indicated? How shall we plant so as to secure in season those favorable influences of soil, temperature, sunlight, and moisture conducive to the development and proper to the nature of the plant at each particular period? Having in view the peculiarities of early growth exhibited by the cane, when should the planting begin?

The true answer to these inquiries, if I mistake not, will be found to be much at variance with the practice ordinarily pursued. That a system of very early planting should be adopted is already indicated in the comparison of the cane with the oat-plant, during corresponding periods, and still more clearly if we compare it with winter wheat, to which it bears almost as close an analogy. The first, or early period of its growth, differs from that of wheat only

(34)

in length. The frequent occurrence of "volunteer cane," that is, cane which has grown from seed dropped accidentally in the ground during the preceding autumn, and has vegetated during that time, or in the earliest days of spring, renders the resemblance to wheat still more close and striking. The young plants which have thus begun to grow in the fall survive the cold of winter, in the climate of Central Missouri and other States of the Southwest, the roots remaining in the ground uninjured, and when the frosts are over they shoot up rapidly, and, if properly cared for, far outstrip in growth and early ripening any plants from seed deposited in the spring. Do not these facts indicate a mode of treatment for the cane as closely resembling that adapted to wheat, as the constitutional tendencies in each are found to be alike? The order of development in both is the same, but they differ in the length of the successive periods. The wheat plant finds a congenial climate, in the early stage of its growth, in the cool weather of late autumn and early spring, and in winter beneath the snow. Says Liebig:

"The action of a low temperature in autumn and winter, which puts a certain limit to the activity of the outer organs, without altogether suppressing it, is essential to the vigorous thriving of winter corn. It is a most favorable condition for future development, if the temperature of the air is below that of the soil, so as to retard for several months the development of the outer plant."

At what period, then, if the analogy holds good, is that "favorable condition" of temperature to be secured for sorghum? Plainly, during very early spring. Although found capable often of surviving the winter in milder climates, it is neither necessary to its habits nor desirable, that, as in the case of winter wheat, the first period of its growth should be extended back beyond the time in early

spring when the seed is capable of being awaked into life.
The germination of the seed of all plants in the open
ground, uniformly occurs in the spring, at a certain point
of temperature, which is the same for all plants of the same
species, but differs with the species. The degree of mean
temperature which determines the germination of the seed
of sorghum is generally reached early in the month of
April, in the climate of the Middle States, and if the soil
has been suitably prepared, the young plant will continue
to grow from that time forward. Growth is maintained
under conditions which would be ruinous to Indian-corn,
yet, reasoning from a false analogy, we have been in the
habit of planting cane at a season when, if the true habits
of the plant had been regarded, the larger part of the
most critical and extended period of its growth would
already have elapsed.

By early spring planting its demands for a large extension of the root are complied with, for in the low temperature prevalent then only, can that full measure of vital
energy be concentrated in the root, which is afterward so
essential to the rapid and perfect development of the stalk.
If the early stage of its life is passed during that time, it
will be prepared to receive at the time of "shooting," when
it needs it, the sudden accession of heat which gives to our
later spring the character, as to temperature, of an English
midsummer.

In the latitude of Philadelphia, 40·5° and 62°. Fahr.
are the measures of average daily temperature, which are
assumed as the limits defining the beginning and end of
the meteorological spring—or the season of leaf expansion in general—commencing with the swelling of the buds,
and the first signs of active vegetation, and ending with
the full expansion of the foliage. It includes a period of
about sixty-five days, or from the 15th of March to the

19th of May. Within these limits we may locate the sixty days of the early growth of cane, in no case allowing them to terminate later than the early part of June. It should be borne in mind, that a period covering almost one-third of the whole life of the plant should have been passed at that time.

By late planting, the season which is naturally devoted to the extension and ramification of the roots is improperly abbreviated at the expense of the vigor of the plant. Before the root has prepared itself for the grand display of vegetative activity that naturally so suddenly succeeds this stage of slow external development, the warm sun of early summer forces the stem to a premature growth. Lacking the elaborated nourishment which it would otherwise have received, the stalk grows up weak and slender, and imperfect in its structure and functions. Plainly, then, we have been guilty of a great error in sanctioning a system of culture so much at variance with the nature and real wants of this plant. A good crop of oats could not reasonably be expected to follow early summer sowing, and it is a testimony rather to the remarkable hardihood of the plant than a compliment to our sagacity or skill, that a remunerative yield of cane ever follows similar treatment. Let us make it a rule to plant as near the beginning of April as the condition of the weather and the soil will permit; and to render it practicable ordinarily to do so, fall plowing and ridging are necessary. Frosts may cut down the young blades to the surface of the ground, but the vital parts will be uninjured, and when the proper season comes, the superstructure of stem, leaf, and flower will be carried up with a vigor that will evidence the value of the underground foundation from which it springs. No plant is more beautifully adapted than sorghum to the peculiarities of our climate.

4

Early planting, it may be said briefly, is advisable for several reasons.

1. It is conformable to a habit of growth manifested not only by the cane, but by several of the kindred cultivated grasses, as has already been shown.

2. Ripening is secured at a much earlier period by this practice than by late planting. It is reached at a time when the days are long, and the weather favorable to the process of sugar manufacture.

3. Early planting necessitates fall plowing, and assigns the work of preparing the soil to a season of comparative leisure when it can be well done.

4. The germination of the seed is permitted to take place at a congenial period. A season of prolonged drought often commences late in the spring, which either retards or entirely prevents vegetation. This never occurs in early spring.

5. Gypsum, a highly valuable manure to the young cane, is but sparingly soluble in water (1 part in 400), and as it acts only in the soluble state a large supply of moisture is necessary to the exercise of its proper influence. In the wet weather common in early spring its full effect will be produced, but in dry weather it will exert no influence whatever.

6. The fact that early spring planting only is possible in northern China, the original source of the "Chinese cane" now grown in this country, strongly indicates the propriety of adopting the same practice here. (See Ch. 30.)

FALL PLOWING AND DEEP TILLAGE.

As the period of profitable tillage is comparatively short after the young cane plants appear above ground, it should be improved to the utmost while it lasts, and the preparation of the soil for planting should be as perfect as possible. Deep plowing and thorough pulverization at a time when such work can be thoroughly done, will atone for some defects in the after-culture when less leisure and opportunity are afforded; but no subsequent efforts can compensate for the check sustained by the young plants in shallowly plowed and cloddy ground. By a proper mechanical condition only of the soil can its chemical constituents be rendered available to the root. The roots of the cane are long, fibrous, and minutely ramified, in a deep soil, and the rich and rapid development of the stem is less to be wondered at when it is considered that, into it as a common reservoir, these myriad underground conduits pour all at once their combined streams when stimulated to activity by the early summer sun at the proper time. Deep tillage lies at the foundation of all improvement in modern agriculture, and in cane culture especially it should engage far more attention than it has hitherto received. Says Prof. S. W. Johnson: "It is obvious that other things being equal, the deeper the soil the more space the roots of crops have in which to extend themselves, and the more food lies at their disposal. By deep culture new fields are discovered beneath the old, and it is possible to realize the apparent absurdity of more land to the acre." It is not too much to say that by this means the crop of cane usually obtained by shallow tillage may often be doubled.

As connected with deep culture, *fall plowing* is highly important; to early planting it is indispensable.

1. When this plan is practiced, immediate advantage can be taken of favorable weather for planting, when the proper time has arrived.

2. The work can be better done during a period of comparative leisure, in the fall, and when the soil is in good

condition for being broken up. It is drier, and not so liable to become baked and lumpy as in spring.

3. It secures the powerful mechanical aid of the frost in producing thorough pulverization, and that light, open, aerated character of soil, which is so admirably adapted to the growth and nutrition of the young rootlets. The assistance thus rendered by the winter frosts is such as can not be gained by the use of any other mechanical means or expenditure of human labor. The constituents of the soil are thus rendered more active by the admission of air and water; they are more thoroughly distributed, and are more accessible to the roots.

4. In sward land it is doubly necessary in promoting rapid decomposition of the sod, and in destroying noxious insects. The double Michigan plow is a most effective implement in burying the sod deeply and covering it with a smooth surface of mellowed soil, but the subsequent "ridging" for planting instead of being deferred until spring, in this case perhaps had better be done immediately after plowing. The ridges will be sufficiently elevated if formed by a light corn plow from the loose overlying stratum of soil left by the double Michigan, inasmuch as efficient drainage will be secured through the half-inverted sod beneath; also the disintegrating action of the frost will be much increased.

CHAPTER V.

Ridging of the Land indispensable when Early Planting is practiced—Mode of Ridging adopted—Enumeration of the Peculiar
Benefits due to this Practice—After-culture of the Cane described—Illustration.

To secure a proper drainage of the soil during the cool
and moist period of early growth, it is necessary to deviate
from the usual mode of planting on a flat surface. The
advantages of early planting can be attained only at that
period by providing a fit receptacle for the growing roots.
However suitable may be the condition of the weather at
this time, an excess of moisture in the soil will defeat the
great object to be gained. Underdrainage is highly beneficial, but this practice is yet too little in vogue to be appreciated, and attended by too much expense at present
to procure its general adoption. Ridging in this case
secures to the young plant its prominent advantages, and
some others during the early growth of cane which it does
not possess.

The ground which has been plowed during the previous
autumn or winter should be leveled with the harrow before
planting time, and the mellow soil afterward thrown into
ridges 3½ to 4 feet apart by the plow. If the work be
done with care, the pulverized earth thrown from the opposite furrows will form a ridge, which will be double-
crested, or broad on the summit with a slight furrow in
the center. Otherwise, the top of the ridge can be leveled

4*　　　　　　　　　(41)

and slightly furrowed by a light shovel plow or marking instrument, but generally this is unnecessary. If the seed be planted on this ridge and lightly covered, growth takes place under peculiar advantages.

1. Drainage during the wet season, when the growth of the young roots is in progress, is secured together with all its most prominent recognized benefits, such as rendering the soil permeable to the rains, which as they fall, rapidly percolate through it, carrying with them to the roots the fertilizing materials that they have collected on their way from the air and the upper layer of the soil, as well as the heat that they have absorbed, and each time followed on their retreat by accessions of fresh air entering through the open pores. The roots are enabled to penetrate more deeply, thereby resisting the injurious effects of drought.

From water-logged lands, on the contrary, fresh air which is as necessary to the root as to the leaf or the breathing animal, is excluded by the cold, standing water which fills the pores of the soil, and the seed rots; or if germination has taken place, the young plant is literally drowned. The warm rains wash the surface of the soil without entering it or conveying any of their heat to the interior, and, when the sun shines, a hard surface crust is formed that is almost as impenetrable to vegetable growth as if it were of ice.

2. By ridging, a more thorough aeration of the soil in contact with the roots is effected; land thrown in ridges exposes a larger surface than when flat.

3. Evaporation being promoted where the surface is ridged, the land is brought much sooner into a cultivable condition after wet weather.

4. The young cane being upon the top of the ridge has the advantage of elevation over weeds that may spring up around it, and grows more rapidly, receiving more light and air.

5. The growth of weeds, however, may be effectually prevented by the superior advantages for early cultivation afforded in this mode of planting, since, by following the ridge, the surface may be stirred by the plow to kill weeds before the cane is up, if necessary.

The time spent in forming the ridge is not greater than in marking out two rows with the plow, according to the common plan, and the additional labor is insignificant in comparison with that which it subsequently saves.

The after-culture of the cane should conform in its general features to the system thus adopted at the outset, and to the relations of the plant to external influences during the different periods of its development. The course described below may, in most cases, be adopted with advantage.

At as early a period after planting the seed as the condition of the soil will permit, a horse-hoe or cultivator, with the blades shortened on the side next the ridge, should be passed between the rows, stirring shallowly the soil of the ridge, as close to the crest as possible without disturbing it, or preferably perhaps a shovel-plow may be run very lightly along the side of the ridge, cutting from its base, without penetrating so deeply as to disturb the earth in contact with the rootlets, thus breaking the crust on the sides of the ridge and exposing the soil near the roots to the warm rays of the sun. A few days afterward a corn-plow should be passed through, this time opening a broad and deep furrow at a greater distance from the ridge than the first that was made when the ridge was formed, throwing toward the cane not only the earth that had been displaced by the last plowing, but also a fresh slice from the undisturbed space between the rows. A thorough hoeing or raking of the soil on the ridge, to destroy any weeds springing up among the cane, and to loosen the surface, should immediately succeed the second plowing. At this time, also, the cane should be thinned out if necessary, so that the stalks may stand at an average distance of not more than 6 to 8 inches apart in the row; or if in hills, 2 feet or $2\frac{1}{2}$ feet apart, with not less than four to six strong stalks in each hill. These will tiller or "sucker" often to twice or thrice their original number, and when early planting is practiced, as above recommended, the side shoots will show scarcely any difference, either in size or in their period of ripening, from the central stalks. Nor are they at all inferior in the saccharine quality of their juice, while the general yield is vastly augmented.

The surface of the soil should be suffered to remain in the condition in which it was left after hoeing until the first period of growth (see page 33) is past, and the roots have begun to extend themselves throughout the portion previously loosened and thrown up by the plow. The shooting forth of the stalk, indicating a sudden reflux of nutritive matter from the root upward, is the signal for a third plowing, by which a fresh portion of earth from between the rows is thrown toward the ridge. If the distance between the summits of the ridges is $3\frac{1}{2}$ feet and the original ridges were 18 inches broad at the base, and two slices of fresh soil of 6 inches each in width were thrown in by the second and third plowings, a vertical section of the ridge will present a slightly rounded outline, with a broad, shallow drain between it and the contiguous ones. The wood-cut represents a vertical cross section of

the ridge as formed at the time of planting the seed, fig. 1, and figs. 2 and 3 respectively show the condition in which it is left after the second and third plowings.

In the warm, mellow soil of a ridge of such form, the growth of the lateral roots is stimulated, as well as of those which strike downward into the subsoil, and the result is that the plants become so firmly anchored as to withstand prostration when full grown, except by the most violent gales.

To secure the greatest benefit from the light and heat of the sun, the ridges should run north and south. Cultivation should cease after the third plowing, if it is made at the time above indicated ; the cane soon takes full possession of the ground, shading it and preventing the growth of weeds. Stirring the soil after the stalk has begun to shoot up rapidly, has the effect of retarding the ripening process and of impairing the quality of the juice.

The objection commonly urged against the ridging of land for corn, does not apply to this mode of treating the cane, inasmuch as undue evaporation during the heats of summer is prevented by the condition in which the surface of the cane field is left at that time.

CHAPTER VI.

MANURES.

The Food of the Plant—Special Manures—Influence of Manures
rich in Nitrogen at the various Stages of the development of
Cane—Analysis of the Ash—Function of the Mineral Food or
Ash-ingredients—The Elements of Sugar, whence derived—Precautions necessary to be observed—By what Means Exhaustion
of the Soil by Successive Crops of Cane may be prevented, without practicing a System of Rotation of Crops.

IT has been fully proved by experiment that the perfection of saccharine richness in the cane can be attained
and preserved only by care in providing for the plant suitable food. The fact that certain kinds of soil, differing
from each other only in the relative proportion of some
few chemical constituents, influencing very differently the
plant as to both the quantity and the quality of its saccharine secretions, indicates the propriety of ascertaining definitely the nature of the influence exerted by these ingredients of the soil under various conditions, and of applying to it as fertilizers those substances which, not injurious
in other respects, stimulate the formation of sugar in the
juice. On the other hand, those special manures should be
withheld, which, however beneficial in other respects they
may seem to be, at that period, directly hinder the plant in
the performance of its peculiar function.

The specific action of certain agents upon the constitution of the juice is also variously modified accordingly as
they are applied at different stages of growth and devel-

(46)

opment. Their influence may be beneficial at one period
and highly pernicious at another. Ammonia and manures
containing nitrogen, generally, are good examples of the
truth of this observation.

No fact connected with the culture of either sorghum
or the tropical cane has been more clearly established by
the experience of planters than that manures of animal
origin are not only injurious to the quality of the sugar,
but present great obstacles to its formation when called
into action during the middle and later stages of growth.
They favor the production in the juice of substances containing nitrogen, of which albumen in its various modifications is the common representative. This substance is
always augmented in all plants by giving them a large
supply of ammonia conveyed in the form of animal manure. The gluten of wheat, for example, is commonly
more abundant according to the quantity of such manure
applied to the soil in which wheat is grown. This substance is an essential constituent likewise of the seed of
cane. A certain amount of nitrogenous matter in a mucilaginous form necessarily exists in the juice previously to
the ripening of the seed, and at that time a sufficient quantity of it forsakes the stem and leaves, and in the form of
gluten, etc. becomes fixed in the seed. At the period of
ripening, therefore, the quantity of azotized mucilage or
albumen previously found associated with the sugar in the
juice, is diminished in proportion to the quantity consumed
by the seed. The juice is purer then than at any former
period. An increase in the quantity of the seeds also follows an increased supply of ammonia.

In general terms, therefore, it may be asserted that ammonia applied as a special manure, in large quantity, is
important when a large quantity of *grain*, abounding in
nutritious substances, is the main object of the cultivator.

On the contrary, when the perfection of the saccharine
matter in the juice is the desideratum, the supply of ammoniacal manures, during the period of rapid growth, should
not be increased beyond the amount which the plant always naturally derives from the soil and air. When this
precaution is not observed, a large but weak and watery
stem shoots up, the juice of which abounds in albumen and
impurities, is insipid to the taste, deficient entirely in
crystallizable sugar, ripening late or not at all, easily decomposed by frost, and comparatively worthless. Cane
grown under such conditions is forced into an apparently
luxuriant but really imperfect development; it is more liable to disease, and the stem, lacking the stiffness necessary
to sustain its weight, seldom escapes prostration by the
winds.

In ammonia, therefore, we have an agent, the undue use
of which would soon reduce the sugar cane to a condition
not superior as to the quality of its juice, to that of broom
corn. Happily, however, there are other substances, which,
as we shall see further on, exert a no less decided antago-
nistic influence, promoting the formation of sugar and
encouraging a healthy growth.

Yet there is a period at which experience has proved
that animal manure in very limited quantity may be applied with advantage to cane; namely, during that stage
of its existence immediately succeeding germination;—but
it should be applied only as a top-dressing, or so sparingly
as not to influence the process of sugar formation when the
stem has grown out, and the cells are being stored with
juice. During active vegetation the bulk of the nitrogenous (albuminous) matter is found in the young and
rapidly growing parts,—when not supplied in too great
quantity it is attracted toward those parts exclusively. It
plays a most important part in inducing those changes in

the non-nitrogenous elements by which the vegetable structure is built up, and sugar is finally produced in the ripened juice. Its work is where growth is in progress: in the midst of continual transformation it remains unchanged; the scene of its activity is constantly shifted; as soon as development is perfected in one part it deserts it for another, and hence it is that but a small percentage of it is necessary for the discharge of its legitimate function,—so small that the great bulk of it is found at last lodged in the form of gluten in the seed. A tendency to continued growth of the cellular structure without a proper elaboration of its contents marks the presence of an excess of nitrogenous food; hence the reason why the upper part of the stem of the Southern cane in Louisiana never attains maturity, and is always discarded. So too in the but partially ripened sorghum, albuminous matter abounds in the yet growing upper joints, which it does not desert until the seed is fully mature.

2. Ammonia not only enters into the composition of the juice as an organized compound, but generally also may be detected in it in the form of a neutral salt, and exerts a pernicious influence. This neutral salt is a compound of ammonia with probably oxalic acid. Liebig found it even in the comparatively pure juice of the sugar maple, and it is a source of great loss and injury to the sugar manufactured from the beet. It becomes decomposed by the heat, and the ammonia is volatilized along with the steam diffusing its powerful odor around; the neutral salt by its loss exhibits acid properties, and by the free acid thus formed, a part of the sugar is converted into uncrystallizable sugar. In some samples of sorghum grown upon land enriched with barn-yard manure, not only the peculiar ammoniacal smell is emitted on evaporation, but the taste is equally char-

acteristic and nauseous. Not even a tolerable syrup can
be made of it in the ordinary way.

3. The deleterious influence of nitrogenous manures
upon sugar-producing plants, is to be ascribed, in part,
also to undue stimulation, under the power of which the
plants absorb not only more food than they can assimi-
late, but other substances not their proper food. The
ammonia enters into combination also with the fixed mineral ingredients in the soil, which are thus largely absorbed.
Thus by the action of these manures many salts are introduced into the circulation, which are not only foreign to
the plant in its normal condition, but highly injurious to
the constitution of the sugar. Among these the chlorides
of sodium and potassium, and probably the sulphates, are
said sometimes to be introduced into the juices of the
Louisiana cane in sufficient quantity to impart to them the
peculiar sensible qualities of those salts. Southern planters
are well acquainted with the fact that the juice of canes
grown upon new or recently cleared lands is almost uncrystallizable (McCulloh, p. 580), and the same is true of
sorghum grown under like conditions in the North. The
cause seems to be the great absorptive power or avidity of
the humus in such a soil for ammonia which it derives
from the air. The ashes of the forest which previously
covered it, furnish a superabundance of salts which the ammonia renders capable of speedy absorption. In soils from
which the chloride of sodium or common salt is extracted,
it imparts the property of deliquescence to the sugar, rendering crystallization extremely difficult and destroying
the crystals after they are formed. Experiments with
sorghum have been reported in which these and other
qualities resulting from high or overfeeding, have been
so predominant as to render the syrup in its usual raw
state disagreeable to the taste and worthless for sugar.

These fertilizers, therefore, should be applied sparingly. The elements which they contain should be applied in proportion to the *wants* of the plant. During the earlier stages of its growth, it might be stimulated to greater activity with advantage. And for this purpose the application in small quantities of Peruvian guano or hen manure may be recommended. This, in addition to the ammoniacal salts, contains the different phosphates in large proportion, which are important constituents of the cane, but it should be applied only in such quantity as that its highest stimulating power may be exhausted before the plant has attained sufficient size to absorb more pernicious ingredients. Although but a small percentage of the materials for the growth of the cane is derived from the soil (6 per cent. Lousiana cane), the elements of sugar, carbon, and water being obtained from the air or the decay of plants in the soil, it is well known that the mineral ingredients or salts appropriated from the soil are indispensable to the plant. The following analysis of the ashes of the different parts of the plant will show of what these consist. They were made with the utmost care, from fresh and ripe specimens of a good quality, by Dr. Jackson, of Boston. (Ag. Rep. Patent Office, 1857, p. 192.)

Of the *seed*, 1000 grains on burning gave 27·8 grains of gray ashes, composed of

Silica	10·000	grains.
Phosphoric Acid	6·740	"
Lime	0·200	"
Magnesia	3·580	"
Potash	4·060	"
Soda	2·270	"
Chlorine	0·018	"
Sulphuric Acid	0·222	"
Carbonic Acid	0·600	"
Oxide of Iron and Manganese, with loss.	0·110	"
	27·800	

Of the *dried plant without the seed*, 5·359 grains, gave 205 grains of gray ashes, which yielded on analysis:

Silica	85·854	grains.
Phosphoric Acid	18·245	"
Lime	33·986	"
Magnesia	2·870	"
Peroxides of Iron and Manganese	3·034	"
Potash	30·358	"
Soda	14·534	"
Chlorine	1·693	"
Sulphuric Acid	7·702	"
Carbonic Acid	6·560	"
	204·836	
Loss	164	
	205·000	

The relative amounts of these constituents, it should be noted, are not proportioned to the degree of their importance respectively in the economy of the plant.* Those

---

* "The Prince Salm Hortsmar, of Brunswick, has made the function of the mineral food of the plant the subject of a most extended and laborious investigation. In experiments with the oat, he found that when silica was absent from the soil, everything else being supplied, the plant remained smooth, pale, dwarfed, and prostrate.

"Without lime the plant died in the second leaf.

"Without potash or soda it reached a height of about three inches.

"Without magnesia it was very weak and prostrate.

"Without phosphoric acid it remained very weak, but erect, and of a normal figure, bearing fruit.

"Without sulphuric acid it was still weaker, was erect, and of normal figure, but without fruit.

"Without iron it was very weak, pale, and disproportioned.

"Without manganese it did not attain perfect development, and bore but few flowers.

"Other experiments proved that chlorine is essential to the growth of wheat."

He also "found that oats grown with the addition of fixed mineral matters (ash ingredients) only, gave four times the mass of vegetable matter that was obtained when these were withheld."—*Lectures on Ag. Chemistry at the Smithsonian Institution, by Prof. Samuel W. Johnson.* (Smith. Rep. 1859, pp. 130–31.)

which are appropriated most largely are not more indispensable than others which appear in comparatively minute proportions. These ingredients are present in almost every soil in sufficient abundance, when they have not been removed by an improvident system of cropping. As sugar is composed of elements derived entirely from the air, it is evident that as in the case of the Louisiana cane, there would be no necessity for rotation if *all* that has been taken from the soil is returned in the *trash*. A soil which possessed these ingredients at first would not, by successive crops of cane, become exhausted at all, if the *sugar only* were taken away. Practically, however, this cannot be effectually accomplished. The uncrystallizable portion of the juice contains salts which are annually removed with the molasses that is marketed. These are chiefly the *phosphates of lime* and *potash* and the *carbonate of lime*. The leaves are used for fodder, and much is usually wasted before their conversion into manure. The seed, also, if not returned to the soil upon which it grew, abstracts in large quantity the same and other ingredients. Hence, while it may be asserted in general terms that the begasse or trash is a most valuable manure when applied to the cane, it will be necessary to return from other sources the equivalent of this annual waste.

5*

CHAPTER VII.

MANURES (CONTINUED).

The Begasse or Cane Trash—How it may be disposed of—Uses—
Its Composition—How to convert it into a most Valuable Manure—Rationale of the decomposition of the Trash—Waste
Products—Their Use as Manures—Mineral Manures, from Artificial Sources, how supplied—Gypsum as a Special Fertilizer—
Harris's Experiments—Analogies and Conclusions—Mode of action of Gypsum as applied to Cane—Why most Energetic when
applied to Early-planted Cane—Importance of Potash as a Fertilizer.

MOST persons, who have worked up large crops of cane,
have experienced considerable difficulty in converting the
trash into such a condition as that it may be returned to
the soil in the form of manure. The trash, as it comes
from the mill, has generally been allowed to accumulate in
a loose pile, in which condition, in fair weather, it soon
dries, and it is almost completely indestructible by the ordinary action of the air. This property of the dried begasse
suggests the propriety of using it for thatching. As a
covering for sheds, in which the canes are stored, in places
where lumber is scarce, it would no doubt be a valuable
material. In some places, it has been disposed of to the
paper manufacturers. It may be used in the production
of a valuable red dye; or it may be employed for fuel, in
the fresh state, by means of a suitable arrangement of the
furnace for evaporating the juice; or the greater part of

(54)

it may be burned by setting fire to the stack which, not to speak of the risk of destroying other property, is by no means an economical plan.

A much better manner of disposing of it would be to convert it into manure in the barn-yard. This will be all the more readily done when the mill is placed upon a platform above the yard.*

The decomposition of the trash is promoted in a higher degree in the barn-yard than elsewhere. When dry, the great bulk of it consists of woody fiber, some sugar and salts, constituting about 30 to 35 per cent. of its original weight; but in its ordinary condition more than half of its whole weight is juice, which the mill has failed to extract. It is a well ascertained fact, that there are some substances which are in a high degree promotive of the decomposition of woody fiber by enabling it to absorb oxygen. Others, in an equal degree, retard this process. Of the former class are alkalies, and of the latter, acids. When the fresh trash is suffered to lie in a loose heap, fermentation of the juice which it contains, almost immediately sets in, but the acetic acid then formed becomes concentrated by evaporation, and this, with the absence of sufficient moisture, prevents the speedy decay of the woody and cellular matter. On the other hand, alkalies, by inducing oxidation, rapidly promote decomposition when assisted by an elevated temperature, and free access of air, and the presence of a sufficient amount of moisture. Hence, in a barn-yard, where the stable manures, and all the refuse of a farm, the ashes, etc., are mingled with the cane trash spread in a broad bed over the yard, the alkalies which they contain come in contact with it. Where evaporation is checked, and an

---

* Refer to Ch. XXV.

elevated temperature is maintained by fermentation in a
sufficiently compact mass of the trash, and, at the same
time, where enough of air has access to it to permit the
process of decomposition to go on, and the resulting carbonic acid (which arrests the process by its presence)
readily to escape—we have all the conditions supplied
that are essential to rapid decay. The trampling of cattle
over the mass will still further promote decomposition by
the mixing of the ingredients, and rendering the layers
more compact, which is especially necessary at an early
stage of the process.

A still greater improvement consists in first spreading a
layer of trash upon the floor of the yard to a depth of a
foot and a half; covering this with a layer of muck, loam,
or clay, and then spreading the rest of the trash as fast as
it accumulates, and as compactly as possible upon the
muck—the whole finally covered with stable manure. The
lower layer of trash serves to conduct away the superfluous
moisture that, during rain, percolates through the mass—
the muck or clay retains the alkalies, and prevents their
loss, and adds an important ingredient to the compost pile
when it is ready to be hauled away.

All waste or residual products resulting during the
manufacture of sugar or syrup, should be carefully preserved and added to the manure heap. The precipitate
which falls in the clarifying tank, or which becomes incorporated with the green scum in the evaporator, when sulphate of alumina is used, largely consists of gypsum or
sulphate of lime, which is equal, bulk for bulk, as a fertilizer, to the commercial article. The washings of the
charcoal filters—charcoal dust, the ashes from the furnace
and from other sources—should all be incorporated with the
decomposed trash, and annually carried out upon the fields.

Where there is reason to suspect a deficiency of any particular ash ingredient in the soil, it should be at once supplied. Bone dust is a convenient form in which phosphoric acid may be furnished. Potash and soda are returned in the form of wood ashes. Quick-lime unmixed with animal manure should be sparingly but frequently applied. Not one of the least important advantages resulting from the moderate use of lime is the fact that it accelerates greatly the development of the cane, enabling the plant to mature ten to fourteen days earlier than it would otherwise do.

But as a special fertilizer, gypsum must be allowed a most important place. In 1862, Mr. Harris, of the *Genesee Farmer*, made some experiments on sorghum with various manures, in which the advantages resulting from the use of plaster of Paris (sulphate of lime or gypsum) were manifested in a most extraordinary degree. A condensed account of these experiments was published in the *American Agriculturist* (vol. xxi. page 361), from which the following is an extract:

"The soil, a sandy loam, had been under cultivation without manuring for some thirty years; the last three years it had lain in grass and clover. It was plowed and harrowed into mellow condition, and the sorghum planted June 4th, in hills, about 3 feet, 4 inches apart. Eleven plots, each containing one-twentieth of an acre, were experimented upon. The various manures were applied in the hill, being thoroughly worked into the soil, and then covered with fresh earth, on which the seed was planted.

"The sorghum was cut October 7th and 8th, and the stalks accurately weighed in the field, with the following results:

No.	Manure used per acre.	No. of hills.	Actual yield per acre.	Yield of equal No. of hills.
1.	No manure..	128	1,917	3,044
2.	400 lbs. sulphate of ammonia...........	122	9,385	15.387
3.	400 lbs. superphosphate of lime..........	184	21,211	23,059
4.	{ 400 lbs. superphosphate of lime... } { 400 lbs. sulphate of ammonia..... }	142	15,097	21,263
5.	250 lbs. plaster (sulphate of lime).....	184	22,848	24,836
6.	2000 lbs. unleached hard wood ashes	162	13,058	16,120
7.	{ 2000 lbs. unleached ashes.......... } { 250 lbs. plaster (mixed together). }	169	16,105	19,060
8.	200 lbs. common salt......................	161	7,570	9,403
9.	{ 400 lbs. sulphate of ammonia..... } { 400 lbs. superphosphate of lime.. } { 2000 lbs. unleached ashes........... }	112	10,528	18,790
10.	{ 400 lbs. superphosphate of lime.. } { 2000 lbs. unleached ashes........... }	176	21,369	17,920
11.	No manure..	171	3,136	3,662

"Each plot contained 201 hills, but, as the above table shows, many failed. The seed either rotted in the ground or was injured by the manure. The last column shows in the plainest manner the relative yield per acre, allowing the same number of hills to have produced on each plot.

"The effect of plaster (gypsum, or sulphate of lime) is interesting and instructive. Not only does the plot manured with this show the greatest yield per hill, but with one exception (plot No. 3) the greatest number of hills germinated and grew. These two plots, Nos. 3 and 5, were superior to any others, all through the season.

"The superphosphate used was a superior article, made from calcined bones, expressly for the experiment. It should be understood that the best superphosphate contains at least 50 per cent. of *plaster;* so that if common plaster contains 80 per cent. of sulphate of lime, the 250 lbs. applied to plot No. 5 would contain the same quantity of real plaster as the 400 lbs. of superphosphate ap-

plied to plot No. 3 ; if 90 per cent., it would get 25 lbs.
more plaster than plot No. 3.

" The plots receiving plaster and superphosphate are
the two best of the series. Plot No. 5. is a little the best,
and probably received a little more real plaster than No. 3.
One thing is clear : *the soluble phosphate of lime in the
superphosphate* did no good, for on the plot No. 5 we
have plaster alone ; and on the other plot (No. 3) we have
plaster and *soluble phosphate ;* and yet the crop is no
better from the two together than from the plaster alone."

Other experiments made since, render it evident that if the
augmentation of the yield is less striking in some instances
than in the above, the application of gypsum is uniformly in
a very high degree beneficial. Its effect is most manifest
upon soils not rich in vegetable matter, and especially upon
clay lands. Such soils being well adapted to sugar production from this cane, its effects are very marked upon them.

But it is the direct and peculiar influence which it exerts
upon the growth of the plant that is especially noteworthy,
and which, when more generally known, is likely to en-
hance its value as a manure for the sorghum. This consists in the fact that gypsum stimulates some organs of the
plant to extraordinary development, while it has no such
action upon others,—it *increases greatly the size of the
stems,*—in a slight degree the quantity of the leaves, while
it really diminishes the quantity of flowers and seeds. In
the cane, this strong growth of stems is not, as when the
plant is encouraged to a luxuriant expansion by animal
manures, or a rich vegetable soil, attended by any diminu-
tion of the saccharine quality of the juice, but rather an
increased richness. Upon clover, it has long been known
to exert a similar influence. It increases the weight of the
stems at the expense of the leaves, flowers, and seeds. In
plants, as in animals, special *points* of excellence may be

developed, and may be made eventually hereditary by care in keeping the individuals possessing them, through successive generations, under the influence of the agents by which those special forms or qualities were originally induced. Have we not here, then, in this material, the means of producing, within a single season, an influence which the labors of a century in hybridization directed to the improvement of the same qualities, might not have effected, and the promise also of rendering permanent, and of improving to the highest degree, those qualities when once implanted? If broom corn is but sorghum with the saccharine quality undeveloped in it by reason of its subjection to natural influences constantly exerted, may we not apprehend a slow but sure deterioration of the best varieties of the latter that we have, unless some such powerful agent as this is laid hold of to check the tendency to gradual debasement? Sulphate of ammonia has been found to exert an influence very similar to that of the sulphate of lime upon clover.

The mode in which the gypsum acts in producing this change is not understood. Liebig, after many years of assiduous research, expresses his inability to answer satisfactorily this question. He regards its action as very complex. He has lately found, however, that it promotes in a remarkable degree the distribution of potash and magnesia in the soil, by the substitution of those bases for the lime in the combination with sulphuric acid, thereby making them soluble, and rendering "nutritive elements accessible to and available for the clover (at least) which were not so before" (*Liebig,* "Nat. Laws of Husbandry," p. 327). That it also fixes ammonia is certain. Prof. S. W. Johnson estimates that its presence in the juice of plants has the effect of checking undue evaporation from the leaves, thus enabling them to resist drought in a high degree.

However these or other discoveries may indicate its mode of action, the effect is no less striking and peculiar.

Gypsum is best applied to the cane, sown in the row, before the seed, or afterward used as a top-dressing. As *an agent in fixing ammonia*, chloride of lime is probably more effective, being more soluble.

The effect of gypsum will no doubt be most marked on early planted cane, a larger quantity being dissolved by the rains. On cane late planted, and preceding a drought, it would exert little benefit.

Potash is observed to be a preponderating ingredient in the ash of all sugar-bearing plants. It is, therefore, probably not only beneficial directly applied to the plant in the form of wood ashes, etc., but the sulphate of lime, as above mentioned, seems to effect the conversion of salts of potash existing in the soil in an insoluble form, into one in which they directly afford nutriment to the plant.

CHAPTER VIII.

THE RELATIVE VALUE OF DIFFERENT SOILS IN SUGAR PRODUCTION.

The Soil not merely a Repository of the Mineral Food of Plants—Theory alone not a safe Guide in estimating the Value of Soils for different Purposes—Texture and Physical Properties of Soils—The Peculiar Qualities, physical and chemical, of Lands which Experience has proved to be the best adapted to this Cane—The "Bluff Lands" of the Missouri and Mississippi Valleys—Sugar Production in the West and in Louisiana—Soils of the Ohio Terraces.

ALTHOUGH it is true that in order to preserve the existence of every plant in its normal condition, its ash-constituents must be found in the soil in which it grows, it would be false to presume that to insure fertility, it will be necessary only to preserve in the soil a constant supply of each ingredient detected by analysis in the ash of the plant. Science, as yet, has given no satisfactory explanation of the nature of those transformations by which mineral substances in the soil are made to contribute to the life of vegetation. Within the soil, as well as in the organic structure, many subtle combinations are known to be formed which the chemist is unable to repeat in the laboratory. Unexpected results, incapable of explanation according to any known principles, are often developed, and hence the necessity of admitting no inferences in agriculture as facts which have not been referred to the test of experience for confirmation. The relative value of different soils for

(62)

sugar production can only thus be determined. So, too, the specific influence of any fertilizing agent upon the growth or nutrition of the cane, may be regarded as accurately fixed only under certain conditions.

The texture and other physical properties of soils have generally a no less important influence upon their fertility than those that are strictly chemical. The former being readily perceptible are well known, and may be pretty clearly defined; and as they are dependent upon, and, to a certain degree, indicate the chemical composition, they mark the most characteristic points of distinction between different soils.

Correspondent to the quality of the soil, the juice of the cane has been found to undergo, uniformly, various degrees of modification, and although much is yet to be learned on this head experimentally, enough is known to enable us to indicate certain soils as peculiarly favorable to the production of the saccharine matter in the juice of sorghum, almost free from impurities which hinder crystallization, while others are in an equal degree unfavorable in this respect. This difference in lands not artificially manured, appears the most broadly marked when we compare a light, friable limestone soil with a wet alluvial one, or one but recently cleared, and composed in great part of humus or decaying vegetable matter. Upon the last-mentioned variety of soil sorghum grows rank and tall, but its period of growth is inordinately lengthened, the seed fails to mature early, the juices are diluted and mucilaginous, the crystallization of the sugar is accomplished with greater difficulty, and the yield either in sugar or refined syrup is less even than that usually obtained from half worn-out upland soils. The worst possible condition of the juice is attained when the humus is replaced or reinforced by barn-yard or other stimulating ammoniacal manures in large quantity; in such case the cane seldom ripens, and yields about one gallon to twelve of juice, of a dark, pungent, uncrystallizable syrup. Upon such a soil the maize attains to uncommon luxuriance and fruitfulness; and it is this peculiarity which forms one of the most important points of distinction between it and the cane. Upon lands which have ceased to produce remunerative crops of corn, sorghum often thrives well; because, like the pea, clover, etc., it sends its roots down deeply and draws much of its nourishment from the subsoil which is rich in mineral food that the less penetrative roots of the corn have failed to extract.

But if a light, calcareous loam be selected, which is deep and mellow, either comparatively new land which has been long enough in cultivation to have had its vegetable matter thoroughly incorporated with the deeper layers of soil —or old limestone land which has not been suffered to deteriorate by injudicious cropping, and to which no more animal manure has been applied than is necessary to give a stimulus to the plant in the early stage of its growth, the cane will attain to a very large size, it will mature its seed and juice in all ordinary seasons in the Middle and Western States, producing an abundance of a light-colored and easily crystallizable sugar, and a fine rich syrup. If with these advantages of soil is combined a side-hill exposure to the south, sloping at such an angle as to receive the vertical rays of the sun at noon in midsummer, we will then have supplied all the natural prerequisites for bringing the plant to the highest condition of which it is capable. A good variety of cane, under these circumstances, will seldom fail to produce less than one gallon of syrup or five pounds of sugar, from every five gallons of juice.

Such soils are abundantly distributed throughout the whole district where sugar production would be profitable, and especially on the uplands in the Western States they are very largely developed, where the great limestone series of rocks, from which they derive their chief characters, is found.

Among these Western soils particular prominence should, however, be given to one of a peculiar character, consisting of an extensive bed of a fine, friable, silicious marl of a buff color, which is found to cover to a great depth those picturesque hills on the banks of the Missouri River, and of the Upper Mississippi, called "the Bluffs," from which the deposit itself derives its name. It is most largely developed along the Valley of the Lower Missouri, and extends over wide areas of the elevated country on either side, forming the surface of the richest of the upland prairies. The extraordinary and almost exhaustless fertility of these Bluff Lands, as well as other qualities which distinguish them, will be more readily understood when the nature of the agency to which they evidently owe their origin is considered. There can be no doubt that during a comparatively recent geological period, the era of the mammoth, the mastodon, and the huge primeval beaver, a large part of the Western country, comprising at least the basins of the great rivers, was deeply covered by a fresh water lake of vast dimensions. Upon the floor of the lake an extensive deposit of fine, rich, marly sediment was gradually formed, the remains of which, after the subsequent changes of level and drainage of the country through its present water-courses, are now to be found capping the highlands along the streams. This sedimentary bed has in some instances been found to be not less than 200 feet deep, and perfectly homogeneous throughout; but its average depth on the Missouri River, where it attains its best development, generally does not exceed 100 feet, and it is often much less. The soil formed from it contains a very large proportion of vegetable matter in its natural state, derived

from the forests which clothe it, where undisturbed, with
almost tropical luxuriance. Indian-corn, tobacco, or
hemp, should be grown upon the recently cleared land, for
three or four successive seasons, until the vegetable matter
has been completely decomposed and incorporated with
the soil, and its tendency to promote too rank a growth of
cane subdued. After this preparation, it is difficult to
imagine in what respect it could be improved for sugar
production from this cane. Its chemical composition, texture, and great depth, its evenly rolling surface, topo-
graphical elevation and thorough drainage comprise advantages such as, it may confidently be said, are naturally
possessed in the same degree by no other soil in the Union,
for this purpose.

The drainage of this soil is secured by one of the most
remarkable natural provisions. The whole mass of the
upper portion of the bluff stratum, to a depth of sometimes
twenty feet, is permeated in every direction, but more frequently downward, by innumerable cylindrical channels,
which are evidently cavities formed by the decay of the
roots of trees. The green roots not only of the common
white oak, but even of the poke-weed (*Phytolacca decan-
dra*), have been found at a perpendicular depth of seventeen feet below the surface of the bluff soil.*

Experiment has proved that from 300 to 400 gallons of
syrup to the acre may be made from an average crop of
cane on this soil, without the addition of any special manures.
The expense of cultivation is but little greater than is
necessary for the production of a good crop of corn, and
as it is generally conceded that the fodder and seed annually produced amply compensate the cost of cultivation,

---

* Geological Survey of Missouri. First and Second Reports (Prof. G.
C. Swallow), p 71.

the cost of the production of sugar or syrup will be reduced to the expense solely of hauling and working up the canes.

These will compare favorably with the best results of sugar manufacture from the tropical cane in Louisiana; they much exceed the ordinary profits of cotton growing in Mississippi and Carolina, and these facts require only to be known and appreciated in order to insure the permanent establishment of sugar production from sorghum throughout a vast central area of the continent. The fact that these lands are always most fully developed along the large Western rivers, and that their products are thus brought within easy reach of market by water carriage, is an important consideration additional to the advantages above stated. Next to the bluff or an elevated limestone soil of the first quality, the sandy or gravelly terraces of the Ohio Valley deserve a prominent place. These are the best natural soils; the most unsuitable of all others for cane growing are the alluvial bottoms and wet prairies. Intermediate between the best and the worst are many varieties of soils, which by proper tillage and the use of suitable fertilizers, may be brought to the highest standard of productiveness, if the essential requisite of a congenial climate be not wanting.

CHAPTER IX.

THE INFLUENCE OF CLIMATE.

The Limits of the Distribution of Species defined by Climate—Subtropical Character of the Middle and Western States during the Summer Months—The Limits within which Sorghum may be profitably grown defined—The Summer Isotherm of 70° Fahr.—The Summer Line of 72° Fahr. probably the Northern Limit of the District within which Sugar may be most successfully produced.

WHILE the nature of the soil determines in a great degree the situation in which each individual plant is found to flourish in a state of nature, climate fixes the wider limits of the distribution of each species. The geographical range within which the cultivation of the sorghum is capable of being successfully extended in this country, may now be regarded as pretty accurately settled, at least as respects those varieties which have proved the most valuable. This range is very extensive, but within its limits the soil and situation define much more narrowly the area within which alone sugar will become a staple article of production. The results of the past eight years, properly interpreted, indicate this clearly. The subtropical character of those regions of the Old World from whence the several varieties of the sorghum were originally derived, might perhaps lead to the inference that, in this country, we are attempting to carry the limits of the sugar growing region farther north than analogy of climate would support. But such conclusions have no foundation in fact, for it should be remembered that an extreme, almost tropical temperature is characteristic of the middle latitudes of the United States,

(68)

during three or four months of the year. These months cover the whole period of growth of this cane, and where-ever the high summer temperature is sufficiently protracted, and the requisite conditions of soil and moisture are supplied, its cultivation may be successfully carried, although the mean annual temperature of such a region may fall considerably below that of the native country of the plant. Our common corn, a subtropical plant also, furnishes a good illustration of this unique feature of our summer climate in its capability of being here successfully extended in cultivation up to a latitude unexampled elsewhere. But even Indian-corn may ripen in a climate where the yield is too meager to render it a successful competitor with the hardier grasses, or to reward the toil of the cultivator. The same is true of sorghum. The limit of remunerative production will fall considerable short of the line which marks the northern boundary of its growth. A certain line of equal summer heat quite accurately fixes this limit.

So far as temperature is concerned, the growth of any annual plant may be successfully carried to any geographical limits within which the specific amount of heat necessary to the species is uniformly supplied, during a definite period of time. This period, which measures the whole life of the annual plant from the germination to the ripening of the seed, has been found for the larger varieties of sorghum in those climates where it has uniformly ripened and attained to the fullest perfection, to average from 130 to 140 days. The line of temperature which marks the northern range of the sorghum as a sugar-producing plant on this continent, may be defined to be the summer isotherm of 76° as determined by Blodget in his "Climatology of the United States." This line passes from the east through Southwestern Connecticut, the southeastern extremity of New York, Northern New Jersey, Eastern

and Western Pennsylvania, north of Reading and Pitts-
burg, through Northern Ohio, touching the shore of Lake
Erie, Southern Michigan, including Northern Indiana and
Illinois as it passes around the southern extremity of Lake
Michigan, and thence through Southern Wisconsin and
Minnesota.   North of this line, the cane fails to ripen.
South of it, it ripens generally everywhere when planted
early, but this is rendered more certain, and the growth of
the cane is more vigorous when the summer line of 72° is
reached.   There it attains its fullest development.   The
line of 72° crosses the State of Delaware going westward,
extends through Central Maryland into Virginia, where it
passes up the Shenandoah Valley, being sharply deflected
southward by the mountain chains, thence through Western
Virginia, Southern Ohio, Central Indiana, Illinois north
of Springfield, and Southern Iowa.   North of this line, the
period of growth is necessarily extended, and the summers
are gradually shorter and cooler as we approach the line of
70°.   There one hundred and thirty days from the time
of planting are not sufficient, and the necessary amount of
heat which is lacking must be added in September to ripen
the plant.   Beyond this point, the period of growth is extended to one hundred and fifty or one hundred and sixty
days, but before that time has elapsed, the early frosts
suddenly intervene, and cut off the crop.   Occasionally
the same thing occurs far to the southward, but it is happily quite unfrequent.

In the vast extent of country between the line of 72°
and the cotton and cane regions of the South, there is
certainly no climatic obstacle to the successful growth of
sorghum.   The future sugar district of the United States,
instead of being limited to a narrow border along the
Gulf of Mexico, may now be defined to be a great conti-
nental belt of irregular outline, having for its northern
boundary the summer line of 72° Fahr.

CHAPTER X.

MEANS BY WHICH THE MATURITY OF CANE MAY BE HASTENED.

Special Means by which Early Ripening may be secured—Inattention of Planters to this Subject—The Period of Growth may be abbreviated to advantage by the practice of the System of Planting and Culture previously recommended—Selection of the Earliest ripened Seeds from the most Highly-developed Plants—Early Planting—Sprouting of Seed—Drainage of Land—Use of Lime—Upland Soil and Southern Exposure of the Surface—Direction of the Cane Rows—Importance of attention to these Particulars.

To secure for the cane, in whatever latitude it may be grown, the full benefit that it is capable of deriving from the climate and soil, it is important that every part of the work of its cultivation should be done just at the right time, and especially should every means be used by which its maturity may be hastened without injury to its growth, or to the saccharine matter in the juice. It has already been mentioned (in Ch. IV.) that the mode of cultivation there recommended for general adoption, is not only such as seems demanded by the constitution of the plant, but also that it conduces, in a marked degree, to its early maturity. In latitudes north of Washington City and St. Louis, at least, it is an important consideration to secure the ripening of the cane sufficiently early to enable it to escape all danger from frost, and to prevent the annoyance and loss of time resulting from an undue haste in gathering and working up the crop. But for the astonishing neglect of the means con-

(71)

ducive to this end, sugar production at the North would
now be far in advance of its present condition. Not only
would it be found that much time and labor could have
been saved, and the crop brought to maturity much earlier
in districts where the autumnal frosts are not dreaded, but
the richest and best varieties of cane would now be grown
in more extreme situations, with as little risk as in climates
which now seem so congenial to them.

By pursuing the system of planting and cultivation
above recommended, and making use of other means specially directed to this end, the ripening of the Chinese cane
may be hastened from thirty to fifty days. This result
would be more generally appreciated if it were considered
that it is equal to the difference in the length of the season
of places widely separated in latitude, or to the advantage,
in this respect, which the planter in Southern Virginia or
Kentucky enjoys over the farmer in Western New York or
Central Illinois.

This effect is not merely an advancement of the period
of growth, but partly also a real abbreviation of it, and is
the result of a combination of favorable influences, among
the most important of which are a proper selection of the
seed, early planting (or its imperfect substitute—steeping
and sprouting of the seed), drainage of the soil, the application of lime, or the selection of a limestone soil, and
an elevated position with an easy slope southward, the rows
of cane running in the same direction.

An important advantage is gained at the outset in rigidly
observing the rule to plant only the earliest ripened seeds,
or those found upon the summit or at the center of the
panicle and grown upon the largest and best of the earliest
ripened canes. Such seed will produce in turn, not only
an earlier, but a more healthy and robust, growth of cane
than those indiscriminately selected.

Early planting, as already observed, not only accelerates the ripening period, but is in accordance with the habits and natural requirements of the plant. Sprouting of the seed by steeping it in warm water, or in stimulating solutions of niter, chloride of lime, muriate of ammonia, or sulphate of iron, or by pouring boiling water upon it in a sieve, or by burying it in warm soil, or in a hot-bed, is to be recommended only when planting has been so long deferred as to make it necessary, and should never be adopted but in rare exceptional cases.

Drainage, it is well known, is highly conducive to the same result. Well-drained soils are comparatively warm ; the snows melt upon them much sooner, and they are in a condition to be tilled two to four weeks earlier in the spring than those undrained. The soil also is kept in such a condition as to secure to the plant throughout the season the advantages thus gained.

The "warmth" or quickening power of soils in which lime is a predominant ingredient, is universally recognized. "It is true of all our cultivated crops, and especially of those of corn," says Johnston, "that their full growth is attained more speedily when the land is limed, and that they are ready for the harvest from ten to fourteen days earlier." Upon sorghum this effect is equally well marked. This property of a limed or naturally calcareous soil is peculiarly noteworthy, inasmuch as it is upon such land that the cane acquires a richer juice and a more full development of its best qualities than upon any other.

The importance of an elevated situation in hastening the maturity of sorghum has, doubtless, been made evident to every one who has had the opportunity of comparing cane grown on the uplands with that grown in the valleys. The former will uniformly mature much earlier, besides being generally out of reach of the early frosts. High

7

lands produce this influence not only because they are better drained, and less charged with vegetable matter than low grounds, but also for the reason that they are more exposed and airy, and are not bathed in the moist atmosphere productive of rank growth so characteristic of the latter.   The favorable influence of an upland soil in quickening growth is much augmented if it declines gently toward the south, and the effect is further heightened if the cane-rows are made to follow the same direction, thus receiving the vertical rays of the sun at noon, and giving them a more free access to the canes, before and after that hour, than could otherwise be obtained.   It is a question, however, to be decided in each particular case, whether the benefit derived from running the rows north and south, on a surface inclined in the same direction, is not counterbalanced by the injury likely to be done the land by its greater liability to wash in heavy rains, and the danger of prostration of the cane by violent west winds; both of which evils would be avoided by marking out the rows in the opposite direction.

The observance of these precautions would remove the only objections which, in some seasons, might be urged against the general cultivation of the more slowly maturing but richer varieties of cane, in the vast district included between the summer lines of 70° and 72° as already defined.   By these means also, in an extraordinarily backward or cold season, such as is sometimes likely to occur in any climate, a crop may be ripened when otherwise it would be impossible.

CHAPTER XI.

THE "TILLERING" OF SORGHUM.

Influence of the Side Shoots, or "Suckers," upon the Saccharine Products of the Plant—Opposite Opinions of experienced Planters as to the Nature and Value of the Lateral Shoots harmonized—Their Growth under perfect Control—They are not an Abnormal Growth, but are perfectly analogous to the Tillers of Wheat—The Tillering Process common to many of the Grasses, and a wise Provision for their Increase—Under what Conditions the Growth of the Sorghum Tillers should be encouraged.

MUCH diversity of opinion has existed among cultivators as to the nature and influence of the side shoots, or "suckers," which, sometimes to the number of eight or ten, spring up around the base of the central stem. Experiment, it seems, has not hitherto led to the adoption of any well-grounded opinions on this subject. Some planters have been led to believe that these suckers are really what the term, according to the popular interpretation, indicates—a mischievous and abnormal growth, diverting from the main stalk the nutriment which would otherwise have been employed in enlarging the growth and enriching the juice. They advise, therefore, to pull them off whenever they appear. Others have found that these lateral shoots arrive at maturity almost as soon as the central one, are equally as rich in saccharine matter, and of scarcely inferior size, while they vastly augment the yield. They recommend that their growth should be encouraged. So frequently have these opposite conclusions been reached, as to leave no doubt that in each case they were justified by the results.

A careful study of the circumstances under which these results were produced, however, will show that they are in perfect harmony with each other, that they are under perfect control, when the causes producing them are understood, and that they are in obedience to a common law of growth.

The production of side shoots from the base of the central stem of the grasses, instead of being an unusual effect, is one of the most common and efficient means provided by nature for their increase, and even for preserving their existence in their great contest for life with other plants. By this means the blue grass of the meadows by a compact circle of stems and roots is enabled to occupy the soil and to resist the encroachments of other plants in its vicinity, and to the same cause we owe the fruitfulness of the harvests. The tillering process of wheat is perfectly analogous to the so-called "suckering" of sorghum, and, in offering an explanation of the one, we but reiterate facts well known as affecting the other. For convenience, the mode of development may be designated in each case by the same name. It is to be distinguished, however, from a peculiar ramification of the stalk above ground, to which the sorghum is sometimes subject in wet summers. In oats, true branches spring from the axils of the leaves, and these branches bear grain which ripens very unequally, and to this mode of increase that last mentioned corresponds; but true "tillering" is an underground process, and it is characteristic of cane and wheat in common with many other plants of the same family. The lateral stems in this case spring from underground buds. Although thus connected with the central culm, they derive their nourishment from roots of their own, which increase in number and length as the necessities of these secondary stems, to which they are the purveyors, seem to require. The external development of a new side shoot is always simultaneous with the issue of a new set of roots springing directly from it, while the growth of the proper roots of the central stem is not retarded by this process, being more widely extended at that time, and the surrounding earth not being exhausted of nutritive materials. At a later period, however, this exhaustion occurs; but then the widely branching roots seek a fresh supply of food beyond the circle which at first supplied their wants. *An increase of side shoots, therefore, is always an index of an increased development of the root.* Extensive tillering of wheat is always an evidence of vigor and hardihood. Where this does not occur, either the ground is too poor to afford nutrition to new roots to sustain new stems, or the weather during the early growth of the plant has been too mild, and the single central stalk has been pushed forward too fast, the natural period of rest before "shooting" not being long enough to allow of the formation of a more widely extended root surface than is necessary for the supply of a single stem, or it is the result of too thick sowing, the development of the roots being similarly restricted. In all those cases where but a single stalk is produced, it grows up sometimes of average size, but oftener weak and thin, and from the same cause, viz. hinderance of root development. On the other hand, when not too thickly sown, and when the soil and season are favorable to its wants, instead of a solitary stalk and a single meager ear, each seed gives to the grasp of the reaper a clump of healthy stems and nodding heads, producing a hundredfold.

Tillering, therefore, is a natural process, conducive in a high degree to the health and vigor of the plant.

Let us now consider whether the discordant statements

7*

already referred to, as to the effects of this process in the case of the sorghum, may not be easily reconciled, and a rational method of treatment arrived at, adapted to the requirements and true nature of the plant, and to the external conditions of growth to which it is necessarily subject.

It is observed that cane does not tiller when planted very thickly in the drill, nor when it is put in thin land, shallowly plowed or badly tilled ; nor generally when late planted. In the first case mentioned, root growth is checked by too thick planting in the drill ;. the result is that not only tillering is prevented, but the stalks grow thin, soft, and fibrous, instead of strong, hard externally and succulent within. In the next case, where the condition of the soil is at fault on account of natural sterility, or bad tillage, the cane crop is still more a failure. When late planted also, tillering is checked, especially if the season which follows planting is not unusually cool and wet. The cause is plainly due to the fact that the natural period, within which the roots are to be formed, is so much abbreviated that sustenance enough can be gathered only to support a single stem. Has not this system of impoverishment, then, a tendency to degradation? Is it from parentage originally produced by such treatment that we should seek to perpetuate our stock for future use ? Does not the concurrence of these facts lead inevitably to the conclusion that the noblest type of the plant, the healthiest, the richest, the most valuable in all respects, is to be sought in a form exhibiting the freest and fullest development of all its organs in the manner best adapted to its nature ?

The mode by which such a result may be secured is already indicated in alluding to those influences which conduce to the opposite. Deep culture in a light, mellow-well-drained soil, and early planting in hills or in drills, with

the seed so thinly distributed as to permit of each forming a
stool, are plainly indicated. The proper distance between
these stools or hills, in order to produce the heaviest crop
of the best quality of cane, will vary somewhat according
to the time of planting and the nature of the soil and season. But when the above conditions are complied with,
as heavy a crop as the land can ordinarily sustain can be
obtained with the stools about 15 inches asunder in the
row, and the rows 4 feet apart. No suckers should be removed, for, if the proper precautions are observed, no more
suckers will be thrown out than the roots are able to sustain.

It is easy to see that in cases where the growth of side
shoots is encouraged for a time by any mode of treatment
by which the roots are permitted to extend themselves, and
then the development of the latter is suddenly and permanently checked, an immature and badly developed stool
will be the result, with which (thus produced by a mode of
treatment inconsiderately adopted) planters have had good
reason to be dissatisfied. When, on the contrary, favorable influences were secured, especially early planting of
the seed, the heavy yield and excellent character of the
crop, where the suckers have matured, are sufficient to lead
to a very different judgment.

It will be evident from what has been already said, that
the difference between early and late planted sorghum is
similar in kind to that between fall and spring wheat. It
is a difference of growth, vigor, and development. Au-
tumn-sown wheat, during the milder intervals of winter,
when a slow growth is encouraged, acquires a vigorous
constitution; it is allowed sufficient time to develop its
roots, although the blade may have perished in the frost;
and when confirmed mild weather comes, it is ready to put
forth a luxuriant clump of blades and stems. Spring wheat

is not permitted the natural period of rest after germination that the wants of the plant seem to require.   There
is no marked cessation of growth in the blade after it has
appeared above ground; it tillers very little, and requires
to be more thickly sown than winter wheat.   The inferior
quality of the grain of such wheat is generally well known;
the plant is less vigorous; it is not capable of enduring
such vicissitudes as the cold-hardened winter wheat; is
more subject to disease and to some insects, and unless
very thickly sown, is peculiarly liable to injury from
drought in summer.   If we should not desire to perpetuate our stock of wheat from this inferior and unhealthy
sort, is it wise to subject to such an ordeal a plant exhibiting the same characteristics under similar circumstances,
and to expect that, year by year, it will not degenerate?

With the exception that the half dormant period in
sorghum is not so extended as in wheat, and that it has
not sufficient hardihood to endure the cold of winter, the
analogy is perfect.   In point of hardihood it seems to
occupy a middle place between our spring-sown grains and
those sown in the fall.

There are some varieties of imphee it is said which do
not tiller.   If, like that of Indian-corn, this is found to be
their natural mode of growth, it would be unwise to attempt to interfere with it; but if it is the result of the
"cramping" system, as sometimes pursued with the Chinese
variety, we may readily account for the feebleness of growth
and light yield of many varieties of the African cane.

My own observation leads me to believe that the tillering process is natural to all varieties of sorghum. All the
common kinds of imphee grown from seed sown in early
spring, in the latitude of Pittsburg, were found to conform
to it, although the tendency to multiplication in this way
seems, in the case of the Chinese cane, to be peculiarly
strong.

CHAPTER XII.

HARVESTING THE CANE.

Preparation of the Cane for the Operation of Crushing, Removal of the Blades—Their Proper Function discharged at the Period when the Seed is Ripe, when they should be removed—Their Value as Fodder—The Mode of stripping Cane—How and when to top Cane—Removal of the Cane from the Field—Effect of Frost upon Cane under Different Circumstances.

THE preparation of the canes for crushing is best performed in the field. The first step is to strip off the leaves. This should not be commenced until the plant has matured. From the nature and functions of the leaf, as at present understood, we may infer that direct and permanent injury to the whole organism of the plant would be the result of the removal of the leaves before that period.

While they have been compared not inaptly to the lungs of animals, the leaves are known to discharge other complicated functions not analogous to those of the organs of respiration. They are the laboratories in which organic products, such as sugar, starch, and albumen, by complex chemical changes, are prepared for the use of the plant, from the ordinary constituents of the air and soil. They are the outward recipients likewise of very remarkable influences derived from the sunbeam ; namely, those excited by light, heat, and the actinic or chemical rays, all of which reach the inner fabric chiefly through the leaves

The leaves of the cane, however, seem to outlive their

(81)

office. Unlike what commonly occurs, vitality seems to
linger longer in the leaves than in any other organs; long
after the contents of the cells have undergone their final
transformation, and the seeds have matured. Oxidation
does not occur as early as in the leaves of most plants,
and consequently they retain for a time their green color
and vigor.

The proper time, therefore, to remove the leaves is
during the few days immediately subsequent to the ripening
of the seed and the full maturity of the juice. The seed
is then hard, the husk (glume) has received the full depth
of color natural to it (very dark purple or black in the
Chinese cane and Oom-see-a-na, and various shades of
purple, red, and yellow in other varieties), the panicles
(seed heads) begin to spread and droop, and the stalk is
more or less yellow or orange colored, according to the
variety. The blades at that period are almost as green
as in midsummer, and their nutritive qualities as forage
for horses and cattle are then unimpaired, and if properly
cured, they are fully equal, ton for ton, to good hay, and
are eaten with even more avidity. The blades and seed,
if taken care of, will generally repay the whole cost of
cultivation. Of the former, half a ton, and of the latter,
thirty bushels, is an average yield per acre.

In a mature state, the Chinese cane, if not exposed to
severe frosts, may be allowed to remain in the field uncut,
after being stripped, for a few days without injury. Some
of the later ripening imphees, however, will not bear this
treatment, and soon become sour; a condition to which the
Chinese cane, when unripe, is subject under similar circumstances.

The importance of stripping off the blades before the
cane is cut, will be found to consist chiefly in the saving of
a large amount of excellent fodder, and in an economical

use of time, by doing a needful part of the work in advance of the period of cutting, when time could not so conveniently be spared. This operation seems to be attended by no decided influence upon the juice of the stalk, although higher crystallizing powers have been claimed by some for the juice of canes so treated, which are supposed to be due to the more free exposure to the sun and air thus afforded.

The practice of stripping cane by hand in the field is so slow and troublesome as to be impracticable on large plantations, and the labor is still greater after the canes have been cut and piled. Yet this mode has the advantage of securing the preservation of all the fodder in the best condition. But by far the most rapid way is to strike off the blades of the standing cane with a kind of wooden sword, double-edged, about three and a half feet long, and in shape like a long scutching knife, used by flax dressers. With this implement a man can readily blade three-fourths of an acre of thickly set cane in a day. A large proportion of the leaves may subsequently be raked together in heaps, and saved, but almost all that fall in the row are left, as they cannot be caught by the rake either among the standing cane or the stubs. Cattle turned into the fields after the cane has been removed, will eat greedily all that remains, and in some cases it may be more economical to leave all the blades on the ground to be consumed in this way. The rapidity with which cane can be stripped in the way last described, renders inexcusable the slovenly and wasteful practice of passing it unbladed through the mill.

The two upper joints, with the leaves attached to them, should generally be removed along with the head when the stems are topped, and in the operation of stripping, the two upper leaves should not be stricken off. The best topping instrument is a carpenter's broad-axe, the heads being cut off on a block at the time the canes are received from the wagon which conveyed them from the field. Topping cane with a corn knife is wearisome work, and when done in the field renders necessary the additional labor of gathering up the fallen tops, and hauling them in separately. When the two upper blades are not removed, they serve as a guide to the eye of the workman who arranges the cane in bundles upon the block, and the topping is then done with the axe with great rapidity and nicety. If this is done in fair weather, the seed is then in a condition to be put under shelter and dried rapidly, a matter always difficult of accomplishment when the tops have been cut off in the field and thrown upon the ground. The latter, on account of the scanty and impure juice which they contain are valuable only as fodder. After being topped, Chinese sorghum canes will average six to eight feet in length. They are then about two feet longer than the stalks of ribbon cane, grown in the Southern States, when prepared for the mill. On some sugar plantations, not more than three or four feet of the stalk of the Southern cane is suitable for being crushed.* The sorghum stalks are more slender, but they may be grown much more densely on the ground.

---

* De Bow's Resources of the Southern and Western States, p. 269.

CHAPTER XIII.

STORING THE CANE.

The Sugar Factory, what Buildings necessary—Cane Sheds, their Construction and Arrangement—Trash Thatch a Convenient Material for Roofing, etc.—How used—Conditions essential to the Preservation of Cane—Length of Time during which it may be kept uninjured—Experiment—Uniformity of Temperature necessary.

THE cane should be cut and removed from the field if ripe, immediately after the operation of stripping has been finished. If it has been planted early (see Chapter IV.), it will be in a fit condition for removal, ordinarily, before it has been touched by any frost capable of doing it harm. The imphees in general are more sensitive to cold than the Chinese variety at corresponding periods of their growth, but most of them mature earlier, and are not so much exposed late in the season. The fact that ripened cane will usually endure without injury the earliest autumnal frosts, should not induce any one by a system of late planting to run the risk of ruining his crop by exposure to the often intense cold of October nights, a calamity which, by pursuing a different course, might have been entirely avoided.

The damage done to a cane crop by frost is to be measured, however, not by the intensity of the cold, but by the rapidity of the subsequent change of temperature. A common hoar-frost rapidly melting upon cane in the morning sun, does more harm than a ground freeze gradually abated

S   •   (85)

by cool winds and cloudy skies. In freezing, the delicate cell-walls of the plant are ruptured by the expansion of the fluids which they inclose, and the saccharine and nitrogenous principles become intermingled if a thaw suddenly sets in. But if a freeze is succeeded by a gradually increasing temperature prolonged through a period of two or three days, the cane remains comparatively uninjured. The ruptures of the cellular organs gradually close up, and catalysis is prevented. We may easily correct the injurious effects of sudden extremes of temperature by *storing* the canes, after cutting, in a proper manner; but in the field these changes cannot be controlled, and therefore it is indispensable that the crop be taken off the ground and placed under shelter as soon as ripe.

As already indicated, the whole process of sugar production may be conveniently divided into three successive periods. The first of these embraces the period of cultivation and ends with the harvesting and storing of the cane. The sugar-house or factory, however humble, consists essentially of three parts or divisions, correspondent to those periods, and successively used, viz.: 1. A storehouse or repository for the cane. 2. A mill-house and evaporating-room. 3. The sugar-house proper or curing and crystallizing-rooms.

The business of sugar production from sorghum will not be successfully conducted, and will not assume its true rank as a great national industrial pursuit until convenient and substantial buildings are made to take the place of the mere sheds commonly used only to protect the workmen from the elements. Especially when the work is on a scale of considerable magnitude, ample provision should always be made for carrying it forward in each department independently, and in a perfect and thorough manner. Neatness and adherence to system should rule throughout the whole establishment, even in minute details. The buildings should be contiguous and sufficiently roomy.

The cane-house or barn need be nothing more than a large square structure, eight to ten feet in height at the eaves, inclosed on all sides sufficiently to prevent exposure of the cane to the sun, rain, or snow, and covered with a tight roof; or a square area contiguous to the mill, occupied by parallel ranges of long and narrow sheds extending east and west, and screened from the sun at the sides and ends, is a cheap and good substitute. These shed roofs should pitch toward the south, and if they are ranged close together, only the unprotected ends and the south side of the southernmost of the series need be closed up. These buildings may be constructed of any materials that will afford sufficient protection to the cane. A thatch of begasse, the crushed stalks being received from the mill without entanglement, and laid aside in small bundles to dry, would answer an excellent purpose, and would be cheaper perhaps than any other materials, and equally as durable as others much more costly. The same in many places where lumber is difficult to be obtained, as in some parts of our Western country, might be employed in closing up the sides and ends of these buildings just as straw is used in some places for walls of outhouses.* For this purpose the fresh trash will be found preferable to that which has been dried, as it

---

"A convenient way is to set upright poles about eight inches apart, and draw wisps of straw (or trash) round each, so that both ends of each wisp shall be outside. It is best to lay these in horizontal courses, and beat down each course as it is laid, keeping it uniform and tight. As the filling in with straw progresses, there may be a split pole woven in once in three feet or so to hold the uprights in place. The straw is finally to be raked down on the outside so as to shed rain well. This makes a tight, warm, and lasting wall. The inner side is quite even, and it may be sprinkled with mud if there is danger of the animals pulling out the straw to eat."--*Amer. Agriculturist*, vol. xxiii. p. 9.

is more pliable and can be packed more closely. The
employment of such materials would be highly objection-
able in structures close to the evaporating-room or any
parts of the building in which fires are used, but no unusual risk would be incurred if the cane sheds occupy the
proper position, the extremity of the range of buildings
farthest removed from the evaporating-room and sugar-
house.

In such buildings Chinese cane, if not injured by frost
previously to being cut, will keep perfectly uninjured until
January if necessary. The canes having been previously
topped and bladed may be either tied in bundles of forty
or fifty stalks together and set in on the butt ends closely
under shelter, or they may be laid horizontally in compact
piles, and in either case, if screened from the sun and rain,
the density of the juice and the bulkiness of the stored
mass of canes will preserve them almost unaffected by the
sudden fluctuations of temperature common in early winter,
and even if they are frozen by protracted and severe cold
they will thaw gradually in this condition without hurt.
But the knowledge of this fact should not tempt any one
to protract the working up of a crop beyond Christmas.
The confirmed cold weather, which in the latitude of the
Middle States usually sets in shortly after that time, sus-
pends the work. Subsequent evaporation of the juice in the
stalk and repeated freezing and thawing render it almost
worthless before spring. Evaporation goes on as rapidly
in frozen as in unfrozen cane, and is alone a source of great
loss. Experiment proves that no mill is capable of expressing all the juice contained in the stalk, and that when
the juice has become dense by evaporation, a larger proportion of it, and consequently of the saccharine matter, is
retained in the stalk than when it is more dilute. There
is a loss then of sugar or syrup to the manufacturer from

this source directly proportionate to the length of time during which the cane is kept stored beyond the short period of "seasoning" when the juice is in the best condition. The stalk also becomes more tough and fibrous, and thus affords an additional obstacle. Economy therefore dictates the utmost diligence until the work is done.

These facts lead to the inference that the question as to the length of time within which it is possible to keep sorghum cane in good preservation is of more interest to the curious than practically important. Yet an observation directly illustrating this point seems not out of place here. The preservation of the cane seems to depend chiefly upon a single external condition, the temperature of the surrounding air. I have found that it does not undergo any apparent change during the winter months in a uniformly cool atmosphere. The warm intervals which so frequently and suddenly alternate with the cold in our winter climate, often, however, elevate the temperature of the air in the shade sufficiently to permit fermentation to set in. Any means by which these extremes of climate may be moderated and freezing prevented will preserve the cane with but little alteration in the quality of its juice, until natural evaporation has rendered it worthless. Pursuing a plan accidentally discovered, I have kept cane in good condition until the month of March. It was fully ripe when cut, and was laid in small heaps across timbers resting upon the cellar floor of an outbuilding. The door of this cellar was allowed to remain open all winter. The drainage was defective, and water had collected upon the floor to the depth of nearly a foot after the cane had been stored. Cold weather coming on, the cane was not worked up, but I found that the alternate freezing and thawing of this body of water kept the cane above it at a uniform temperature, and *in an unfrozen state throughout the winter.*

8*

This effect may be thus explained. Water in freezing, or in being converted into the solid from the liquid state, evolves heat previously latent, causing an elevation of temperature in bodies in its immediate neighborhood. The cane from its position, in this case, was the recipient of a portion of this heat, and during the whole of the coldest weather when the congelation of the water on the cellar floor below was gradually going on, it was thus kept above the freezing point. The greater density of the juice as well as the protection afforded by the cellular and woody substance of the stalk within which it was inclosed also contributed to this effect. On the other hand, when very mild weather suddenly ensued and the ice on the subjacent water began to melt, it in turn absorbed heat from the cane, but not so rapidly as to reduce it at any time to the freezing point. The surrounding air within the cellar also suffered a depression of temperature in like manner nearly equal to what it had been during the previous cold weather, being thus tempered by the water in the same way that a lake mitigates the extremes of climate on its shores.

CHAPTER XIV.

PROCESS OF MANUFACTURE.

Distinctive Features of the Process of Manufacture herein recommended — Difficulties attendant upon Crystallization from an Impure Solution — Defecating Substances, Requirements to which they must conform — Advantages of a well-defined System — Errors which have hindered Progress in Sugar Production from Sorghum — Necessity of a New Mode of Treatment — Disadvantages of the Common Mode — Constituents of Sorghum Juice — Thorough Defecation the great Desideratum in any Method.

I DESIGN to embody in the following pages the details of a system of manufacturing sugar and syrup from sorghum, adapted to common use. This system is based strictly upon the results of oft-repeated and carefully-conducted experiments, and it will be found to embrace some facts not previously known, the application of some former discoveries to new uses, and the incorporation with these, wherever practicable, of any other modes of working that were found upon trial to be of practical value, and worthy of being perpetuated.

Among the distinctive features of this process are the following :

1. A method by which the crystallizable sugar in the juice is separated from the impurities which prevent its assuming the crystalline form. This consists in the use of a chemical agent which it is believed has not heretofore been applied for the defecation of sorghum juice.

(91)

2. A new mechanical device to effect the separation of the scum containing the green coloring matter and feculancies without the loss of time, the labor, and the inconvenience which has hitherto attended this part of the work.

3. A means of depriving the syrup of the harsh vegetable taste common to it without the use of the expensive and inconvenient appliances regularly employed by the refiner. The syrup is thus cheaply refined at once, during the regular progress of the evaporation.

4. The employment of a finishing pan of such construction and form as to permit the evaporation of the syrup to be protracted to any required degree, without danger of carmelizing the sugar, which may be instantly removed from the action of the heat when the proper point of concentration is reached as indicated by a thermometer, and discharged by dumping it into a cooler—thus affording the means of making readily a pure syrup of perfectly uniform density, or of reducing the syrup to the point of density at which crystallizable sugar may be obtained of the best quality and in the greatest abundance.

The above are inseparable and indispensable parts of this process, and I have found them to be requisite to uniform success in the manufacture either of sorghum sugar, or of a superior quality of syrup. Other topics also, of scarcely minor importance, demand attention. The whole subject is one which is fraught with interest of a higher character than is indicated by the nature and results of the investigations which have come to the knowledge of the public. At the first view, it might seem that the art of extracting sugar from a liquid so rich in it as sorghum juice has proved to be, is not a matter involving any peculiar difficulties in practice; but its apparent simplicity vanishes when it is found that this saccharine liquid contains, intimately associated with the sugar, a variety of other substances of very different chemical properties and relations, some of which are yet but imperfectly understood. Most of these substances, uncrystallizable themselves, prevent, by their presence, the sugar from assuming the crystalline form—the only form in which it can be obtained pure. In order that we may make sugar, therefore, it is necessary to remove first these extraneous substances from the saccharine solution. But the means used for this purpose must be well chosen. They must be adequate; they must be of such a kind as to be adapted to ordinary use on the large scale; the defecating agents must be harmless to health if inadvertently added in excess to the juice, and must leave no harmful compounds in any product which is afterward to be used as an article of diet; they must be sufficiently convenient in form, and low in price, and they must not exert an injurious influence upon the constitution of the sugar. We find ourselves at once confronted by a delicate chemical problem. Here nothing good can be accomplished by hap-hazard. We must have a clear view of the nature of the material to be acted upon; means must be employed commensurate to clearly defined ends; the best methods of using such means must be understood, and a certain degree of dexterity in their use must be acquired. And here, as in all other industrial pursuits when successfully prosecuted, if a strict adherence to *system* is observed, and if the relations of the successive parts of the process to each other are well understood, the different operations will be performed with precision and facility, and what may have caused much perplexity at first, will soon appear perfectly simple.

Much time would have been gained, and many errors in practice would have been avoided, if some important truths had been earlier recognized. One of these is that the juices of these new canes exhibit properties which are different from those of any other plants from which sugar has been extracted. Hence the necessity for a new and special mode of treatment adapted to these peculiarities. Many persons also have been slow to accept the fact that the saccharine matter in ripe sorghum consists almost exclusively of cane or crystallizable sugar, and that previous to the maturity of the plant, this kind of sugar is present, if at all, only in very small quantity—grape or fruit sugar (practically uncrystallizable, and of lower grade) being abundant. The nature of the other constituents of the juice has also been misunderstood. Lack of information upon these and other points, has led some persons whom a little investigation would have taught better, to adopt expensive methods borrowed from the beet and cane sugar manufacture in France and Louisiana, which, in the end, they have been compelled to abandon after much disappointment and loss. The want of success which has attended these experiments, has repressed investigation, and the popular mind, except when excited by some accidental example of a successful crystallization, seems to have gradually adopted the conclusion that the production of sugar requires the use of some incomprehensible and extraordinary means too abstruse and difficult for ordinary practice. And the extreme facility with which a tolerably palatable article of syrup can be prepared for domestic purposes, has increased this apathy, and perhaps more than anything else, checked the progress of discovery. The inventive genius of the country responding to a temporary want, yet not rising above it, has produced a score of evaporators, adapted in a greater or less degree to the production of crude syrup; but these, in the main, embracing only various modifications of a single idea, can serve no higher purpose than that for which they were originally designed. Rapid evaporation, and the constant removal of the scum by skimming, constitute almost solely the basis upon which their efficiency is alleged to rest in making syrup by the ordinary mode; but any one that has made himself acquainted with the nature of sorghum juice is well aware that boiling and skimming, unassisted by other means, cannot be relied upon to produce sugar. The fact that sugar has, in some few instances, been produced by such means, is no disproof of this assertion; for sometimes samples of cane may be selected, the juice in the lower joints of which is almost a pure solution of sugar, and consequently it needs no defecation; but such samples of cane are very rare, and the instances in which sugar has been produced even from them by the use simply of "pan and skimmer" are still more so. I have alluded to these particulars here for the reason that I believe them to embrace the most prominent errors in theory and practice now in vogue among planters of the Northern cane; and in order that the established facts upon which success in the production of sugar from sorghum will be found to be based, may not be confounded with them or misunderstood.

Ripe sorghum juice, as it flows from the mill, is an impure solution of cane sugar. It commonly contains a small proportion of grape or fruit sugar also. Other substances also, in less quantity, are either dissolved or mechanically suspended in the liquid. The chief of these are chlorophyll (or the green coloring matter), starch, dextrine (the gummy matter), vegetable tissue, pectine or vegetable jelly, vegetable albumen, gluten, and casein (the last three containing nitrogen), acids, salts, and extractive matter. Other modifications of these are found in syrup, and are secondary products, and are not contained in the raw juice. Liquid sugar is an example. It is a degraded form of cane sugar, perfectly uncrystallizable, and is produced by the action of heat and imperfect processes of manufacture.

The cane sugar cannot be induced to crystallize while any considerable quantity of the substances with which it is thus associated is present in the solution.   To break this alliance is, then, the great desideratum, and, as already stated, the merits of the system of manufacture hereafter to be described, will be found to consist mainly in the high degree of completeness, ease, and economy with which this is done.   The means used, both chemical and mechanical, have been studiously adapted to each other and to the end in view.

CHAPTER XV.

Description of the Apparatus and the Mode of using it—General Arrangement of the Different Parts—Description of the Evaporating Range, Materials used in its Construction—Tanks, etc.—Details of the Mode of Operation—First Stage—Lime and the Defecating Agent, how employed—Saturation of the Acid—How the Scum is separated without the usual Process of Skimming—Arrangement of Division A—Requisites to secure Success—Precautions—Fuel—Chemical Agents.

IF the reader will refer to the engraving (Frontispiece), and will examine carefully the representation of the evaporating range there given, much unnecessary description of the apparatus which I recommend to be used will be avoided. The relative positions of the crushing mill and the evaporating train may be varied according to circumstances. It is necessary only that the mill be placed upon such an elevation, natural or artificial, that the juice, as it is received from it, may flow into the tank (M), and thence through the descending range of pans into the cooler (F).

The pipe (P) which supplies the tank (M) with juice from the mill, may be cheaply constructed of common chain-pump tubing, except when mills of the largest size are used. This pipe when made (as shown in Fig. 5, Ch. XXX.), consists of a number of lengths of this tubing united together and laid in a trench, extending from the mill to the tank, with two upright pieces of the same mate-

9   (97)

rial jointed to it at the opposite ends.   One of these upright arms is about a foot longer than the other, and its
upper end is fitted into the bottom of a tight box or large
bucket, containing a coarse wire strainer, which is so placed
as to receive the juice as it flows fron the mill.   The other
arm carries a spout at one side, projecting over the side of
the juice tank.   The ends of the main tube which support
these uprights are closed, the one next the mill permanently ;
in the other a molasses gate, or movable plug, is fitted, so
that the pipe which should incline slightly toward this end,
may be completely emptied at any time.   This should always be done at the close of each day's work.   The pipe
should be painted, the joints cemented with white-lead, and
the upright pieces securely braced, and when it is placed in
position, the bucket on the one arm of the tube should be
about a foot above the level of the spout on the other.*

The juice tank M may be made of one and a half inch
pine plank.   The joints are cemented with white-lead, and
it is divided into two compartments by a water-tight partition.   The depth of this tank should not exceed three
feet, but its other dimensions may be extended to conform
to any required capacity.   It is most convenient, however,
to have each half of it of such size as to contain as much
juice as is discharged from the mill in one hour.   It should
be thoroughly painted inside and out, with red-lead or iron
minium, and well dried before being used.

The evaporating apparatus consists of four sections, A,

---

* This is necessary to prevent the overflow of the bucket.   A column
of juice largely commingled with foam, just as it comes from the mill, is
lighter than a column of equal height in the vertical tube at the opposite end from which the foam has separated.   The equilibrium is maintained only by the liquid rising to a higher level in the one arm than in
the other.   If the bucket stands but slightly higher than the discharging
spout it will overflow.

B, C, D, placed in a descending range, the pan A being
the highest in the series.   The first three are stationary;
the fourth, D, is movable on a horizontal axis.   They are
constructed of sheet copper, or iron with wooden sides.
When of the last-mentioned materials, both surfaces are
protected by a heavy coating of iron paint.*   Pans with
bottoms of galvanized iron are furnished to order, but the
zinc coating is not preferable in any respect to either the
plain or the painted iron surface, and is more expensive.
The furnace should be of brick-work, of a uniform inside
width throughout, corresponding to the width of the space
between the abutments in division A.   Each pan of the
series is set perfectly level, but no two of them are in the
same horizontal plane.   The fall from B to C in a range of
medium length should not be less than ten inches; from A to
B and from C to D, two inches each.†   The furnace is suffi-

---

* The advantages of this paint, iron minium, or brownish-red oxide
of iron, "are its solidity, durability, cheapness, and above all, its property of preserving the iron completely from oxidation, and of hardening the wood. . . .   It is destined to supplant red-lead (lead minium)
and other pigments that have been used until now for coating wood,
iron, and other metals.   It forms a smooth and stripeless coat upon the
iron, varnishing as it were the metal and preventing the atmospheric
influences upon the paint. . . .   Iron minium mixes readily with
other colors, such as black, yellow, green, etc., and by so doing a variety
of colors is obtained to the convenience of persons who would not like
the dark-brown of the iron minium paint.   It has been proved by experiments that the iron minium paint lasts twice, and even three times as
long as red-lead paint. . . .   It resists generally the strongest heat.
. . .   It is to be applied in several layers: the first ought to be thin,
the second a little thicker.   It is employed the same as other paints, with
boiled or unboiled linseed oil.   If unboiled oil is used, litharge or any
good siccative may be added, but not turpentine."—(*Condensed from
London Practical Mechanics' Journal in Scientific American*, vol. xi.
p. 387.)

† Printed directions for setting up the apparatus are given on another
page.

ciently deep at the mouth to give it a slight inclination upward from the level of the grate bars, thus securing a good draught; and throwing the flame closely against the bottoms of the more depressed pans, C and D, at the after part of the range. A great saving of fuel and a more uniform distribution of the heat is thereby secured. The division of the evaporating surface into three separate sections, which is indispensable according to this arrangement, is a security against its being warped by heat, which always occurs when a long unbroken surface is used, and the separate sections may be packed for transportation in small space and carried without injury.

The cooler is simply a large shallow wooden box of the shape represented in the drawing, and made impervious to liquids by iron paint. It may be mounted upon a truck, and at the close of each day's boiling may be run along a tram-way into the draining-room, and emptied of its contents.

The evaporating train, tanks, etc. being arranged as described, the freshly expressed juice passes into the pipe through the wire strainer, which separates all the coarser fragments of cane pith, etc., and flows directly into the tank. The outlet of the conducting pipe should be close to the partition in the tank, so that by means of a short, movable gutter or T-shaped spout, the current of juice may be turned into either division of the tank at pleasure. Each division, as already stated, should be capacious enough to hold as much juice to within six inches of the top as the mill is capable of expressing in about one hour. Two cocks or molasses gates (one for each division) must be inserted in the side of the tank within about four inches of the bottom.* The tank should be made of such

---

* Only one of these is represented in the engraving.

dimensions that, with a partition one and a half inches thick, dividing it in the center in the direction of its least diameter, each half will contain just 100 gallons, between the level of a mark made on its side six inches from the top, and the level of the exit cock near its bottom. This is a convenient measure of juice, and can be finished to the "striking" point on one of the larger evaporators in one hour.

Begin by filling one compartment of the tank up to the mark, and then turn the flow of juice into the other. Have ready a tin dipper or measure which will hold about half a gill, mix in a bucket this measure of milk of lime with two or three quarts of juice from the division of the tank already filled, return the mixture to the tank and stir it thoroughly. The lime should be added in sufficient quantity at this time not only to neutralize the small quantity of free acid found in ripe sorghum juice of the best quality, but also to impart to it the very feeble alkalinity which, for a reason to be hereafter mentioned, it should possess at this stage of the evaporation. Juice of a nearly neutral character is seldom obtained in large quantity, and lime will always be needed. To ascertain whether any more lime should be added, dip into the juice already mixed with the milk of lime a small slip of red litmus-paper, or yellow turmeric-paper. No more tempering will be required if the color of the litmus is changed to a tint in which the blue slightly predominates over the red, or the turmeric yellow deepens to a faint shade of brown. But if no such change of color is indicated, more milk of lime must be added gradually until the test paper takes the required tint. Then add to each 100 gallons of the juice one pint of the clarifying solution, and stir thoroughly. The juice is then ready to be admitted into the evaporator. This should be done

while the other division of the tank is receiving its charge
of juice.

Water should previously have been turned into the evaporating range to start with, and a strong fire of dry wood
kindled in the furnace. When the water has begun to boil
in the lower pans, admit the tempered juice from the tank
in a continuous stream. The conduit L will convey it to
the head of the pan A. At the same time open partially
the gate P. The simple and effective manner in which the
clarification is accomplished, and in which the boiling
juice itself is made to do the work of skimming, may then
be seen at a glance. The successful working of this part
of the apparatus, however, is limited by certain conditions
which should here be noted.

From the rear of the pan A, two abutments or ledges
S S extend along its bottom for about two-thirds of its
length, toward the front. This pan projects over the
sides and front of the furnace wall, and the abutments
must be so placed as to stand directly over the inner
face of the side walls, and terminate just over the end
wall of the furnace. That portion of this pan, therefore, which is situated between the abutments is exposed to the full heat of the fire, while the remainder is
comparatively cool. The flow of juice into the pan must
always be duly regulated by the rapidity with which it is
heated, defecated, and finally discharged through the gate P
into the next division or pan B. The bed of juice in the pan
A should be kept at a uniform depth of about half an inch,
and then a sheet of rapidly boiling juice will overspread
the whole central space between the abutments. The surface of the boiling liquid throughout this space is consequently much elevated, while toward the front of the pan,
and within the bays or scum receptacles R R at the sides,
it is not heated. A surface current is created which carries

the scum rapidly forward and down into the recesses R R, where it lodges, and, as it cools, it becomes densely compacted and displaces any unclarified juice that may have collected there at the commencement of the operation. At the same time that the scum is being separated and carried forward, an under-current of cold juice is continually passing onward in the opposite direction and becoming rapidly heated. At a certain point (A) as it advances, it is suddenly thrown into ebullition, when it parts company with the scum, and impelled onward by the colder current behind it, it escapes by the gate P into B. The gates P P, the openings of which are protected by wire gauze, afford at any time a convenient exit also into B, for any clarified juice which may be thrown into the bays when they are not packed with scum. Ordinarily, however, these gates are not used as outlets for any considerable quantity of the juice into B, as there is sometimes danger of unclarified juice passing through them. They should be closed at the commencement of boiling, and afterward opened only just enough to allow the small proportion of boiled juice upon which the scum has floated into the bays, to slowly filter through them. A return current is thus prevented, and the scum is forced back into the head of the bays and remains there. At convenient intervals, once every half hour or more, this dense mass of scum is lifted out bodily by means of a large flat dipper or shovel of such width as to slide closely inside the bays.

A strong heat must be uniformly maintained in the furnace. Feed the fire with regularity, and do not add so much fresh wood at once as at any time to check the boiling in the front pan. If at any time this should occur, close the entrance gate into B, and stop the flow of juice from the tank until vigorous boiling is resumed. When coal can be had, it may be used in small quantity along with the wood; a shovelful thrown in now and then is very useful in quickening the fire, but care must be taken to keep the grate bars from becoming clogged with cinders.

The part played by the chemical agents in this clarification has not been adverted to. It will be noticed on a subsequent page. The most important advantages resulting from their use are not all made apparent at once to the eye; but the visible evidences of purity — transparency, brightness, and lightness of color—are not produced to an equal extent by the use of any other means which I have been able to discover, or that I believe may be innocently employed.

When these clarifying agents cannot be obtained, lime may be substituted for them, but it is far less efficient. A quantity of the cream of lime should be prepared beforehand (Chap. XXX.), of uniform consistence and strength. The exact measure of it which is necessary to be added to a given measure of the juice to cause it to exhibit the faint alkaline reaction required, should be ascertained and noted. Successive batches of the fresh juice may afterward be tempered with the same proportion of lime previously used when it is certain that there is no variation in the quality of the juice. But much caution is here necessary to avoid error, for the degree of acidity is not constant even in the juices of canes taken from the same field, and the means of readily detecting these differences must be at hand, and frequently used. The sense of taste may indeed acquire an acuteness in detecting minute differences of quality in the juice, scarcely inferior to that of chemical reagents; but until it has acquired this delicacy, we must rely upon the test paper alone.

CHAPTER XVI.

PROCESS OF MANUFACTURE (CONTINUED).

Second Stage of the Process—Characteristics of a Refined Syrup—
Mode of Refining the Syrup—Manner of using the Filtering
Drawer—Third Stage—Concentration of the Syrup—The Finish-
ing Pan—Thermometer—The Striking Point as determined by
the Temperature—Other Tests of Density—The Finger Test—
Indications at Successive Stages of Concentration—False Indica-
tions exhibited by Impure Syrups—Precautions necessary in
Barreling Syrups.

THE entrance of the clarified juice into division B, marks
the commencement of the second stage of the process
during which the syrup is further *refined*. In B, the
evaporation progresses with great rapidity. The boiling
juice at the back part of this pan should uniformly indicate
a density of about 15° Beaumé's saccharometer.

An adjustable gate or register R both regulates the concentration to the required degree in this compartment, and
the discharge of the syrup into E,—a shallow cistern or
filtering drawer containing animal charcoal, prepared expressly for this use. Through this filter the hot syrup
passes into C,—the third pan of the series. The effect of
this filtration is the removal of any remaining impurities
which hinder crystallization, and of the extractive matter
not previously separated, which imparts to unrefined
sorghum syrup its peculiar flavor. A syrup prepared by
this mode is strictly a *refined syrup* of the first quality;

at this stage it is perfectly clear, and of the finest color and flavor.

The only attention specially required during this part of the operation is to properly regulate the flow of syrup into the filter E by means of the adjustable gate. It should not be allowed to pass so rapidly as to drain too much the pan B, in which the depth of juice should be at least twice as great as in any other, or before the proper degree of density has been reached. Nor should it be unduly retarded. In either case the filtration will be imperfect.

The filter should be refilled with fresh or revivified bone-black before the beginning of each day's work. If reburnt bone-black is used, the dust should first be sifted out. Take a piece of thin flannel or canvas of an open texture, wet it, and spread it smoothly upon the slatted bottom of the filtering-box. Then fill in the bone-black gradually, packing it down tightly so as to leave no crevices. When it is filled to the proper level, draw a little of the coal from the central parts toward the sides so as to leave the surface slightly elevated at the sides. Cover with a cloth of the same kind as that which lines the bottom, and confine it closely at the edges by means of the wooden frame. Fill the empty space above the cloth with hot water, taking care not to disturb the surface of the bone-black. The stream of water, in pouring it, should be received upon a piece of board or tin plate resting on the cloth. Another mode of filling the filter is to fasten down the cloth on the bottom of the box, close the outlet, and pour in hot water until it is about half full. Then add the bone-black. When this mode is adopted, the bottom cloth must be confined very securely at the sides, or the small fragments of the charcoal will find their way under it into the chamber below.

It is convenient to have a movable stand of the height

of the pan C set close alongside of it upon which to place the filtering-box while it is being filled, and from which it may be readily shifted to its proper position. The pan B should be so placed that the register R, at the back end of it may project its whole breadth over the filter and the forward end of C.

The water should be retained upon the filter until the juice is ready to be let in; then remove the stopper and allow the water to pass out of the filter as the juice enters it. This and the water used in C and D to start with, must be passed out ahead of the juice. At the close of boiling in the evening before the exhausted filter is removed, it should be washed by pouring water upon it through the register-box until the liquid ceases to pass out sweet.

Nothing now remains but to concentrate the syrup to the "striking point." But this must be so done as to preserve the qualities which characterize the syrup after filtration unimpaired to the close of the operation. The subsequent evaporation must be rapid, and the syrup must instantly be removed from the fire when the proper degree of concentration is reached. The pan C and the finishing pan D fully accomplish these ends. After being reduced by boiling in C, it is then passed through a gate into D at intervals, and not in a continuous stream as into the other divisions. The finishing pan is shallow, with a long beak or lip, and may easily be turned upon its axis by means of a lever or cord and pulley as at F, and successive batches of syrup, upon arriving at the proper point of density, are dumped into a shallow tank or cooler at the side of the range. When the pan D is tipped to discharge the concentrated syrup, a damper is drawn which shuts off the flame until the pan is replaced and a fresh portion of syrup admitted.

A standard thermometer, the bulb of which is continually kept immersed in the boiling syrup, indicates its density, and it is by far the simplest and most unerring guide that the sugar boiler can use for this purpose. When sugar is to be produced, the strike should be made at about 232° Fahr. (111° Cent.), but for syrup 228° Fahr. (109° Cent.) is the proper point. By the careful use of this instrument successive batches of the syrup may be brought to the proper point of concentration with perfect uniformity.

There are some other simple tests of the density of boiling syrups which experience has proved to be reliable, and with which every practical sugar boiler should be acquainted. They are convenient at all times in confirming the indications of the thermometer, and in the case of the accidental breakage of the instrument a knowledge of them would be almost indispensable.

The first or lowest degrees of concentration beyond the condition of a very thin syrup are indicated by the appearances that it presents in falling from the thin edge of a skimmer or dipper upon which it has been taken up and exposed to the air for a few moments by turning it about. First, it falls slowly in large drops; then when the concentration is further advanced it separates from the edge of the skimmer in broad thin sheets. Beyond this point the *finger test* or *proof by the touch* is regarded as the most accurate. The following account of it is given by Prof. R. S. McCulloh ("Scientific Investigations in relation to Sugar and Hydrometers," p. 272).

"The *proof by the touch* is used, I believe, alike by the refiners and sugar makers of all nations, both in Europe and America, with very slight modifications. A small portion of the syrup is taken for trial between the index finger and the thumb; when it is cool, the finger is separated from contact with the thumb, and the syrup examined by placing it between the eye and the light. At different degrees of concentration the following indications successively occur:* '1. Two drops separate, that on the thumb and below is the larger. 2. The drops become nearly equal, and do not separate until the fingers are drawn more widely apart. 3. By the separation of half an inch a thread is drawn out which finally breaks below; the end of the thread becomes claviform' (club shaped), 'and it rises slowly toward the finger. 4. The same thing occurs at a greater distance; the end of the thread is folded back, and gives to the thread the shape of a ribbon or long strip, which rises more rapidly than before. 5. After a greater separation of the fingers the thread breaks, being very fine at the end, which turns aside and twists up like a corkscrew. It does not fold itself upon the rest of the thread as before, and the thread does not increase in volume except by the cohesion which draws the particles toward the finger, which is the only adhering point. A little more concentration prevents the thread from shrinking at all upon itself.'

"In the United States the manufacturers of sugar from cane juice use the *proof by the touch*, and consider syrup sufficiently concentrated when the broken thread curls up in the form of a corkscrew, or the fifth of the degrees above described."

It is essential, however, to the correctness of these indications as well as of those of the thermometer, that the syrup be of a standard degree of purity, such as will uniformly result from the juice of ripe cane of good quality subjected to the mode of treatment already indicated. The presence of impurities such as are found in imperfectly defecated syrups, or an excess of either acid or lime, be-

---

* See Dictionnaire de l'Industrie, p. 407. Paris, 1841.

sides injuring the sugar, impair the uniformity of all such indications.   I have found that the fifth degree of concentration, as marked by the proof by the touch above described, is attained in properly refined sorghum syrup when the boiling temperature rises to 235° Fah. (113° Cent.).

If syrup only is designed to be made it should be discharged from the tilting pan when a temperature of 228° Fah. (109° Cent.) is reached. Syrup should remain in the cooler until it has parted with most of its heat before it can be safely transferred to barrels.   If it is poured into the barrels hot, the prolonged action of the heat retained in the dense mass will be sufficient to scorch the syrup.

New barrels should be limed and fumigated with sulphur before they are intended to be used. Old molasses barrels are not fit to contain choice syrup until in addition to this treatment they have first been either steamed or thoroughly soaked with hot water. (See directions for cleansing, Chap. XXX.)

CHAPTER XVII.

PROCESS OF MANUFACTURE (CONTINUED).

Crystallization—Treatment of the Syrup in the Crystallizing-room
—Crystallizing Vessels—Dutrone's Crystallizing Box—How
Drainage may be secured and Crystallization promoted.

AT the close of the day's boiling, transfer the cooler to
the crystallizing-room. Here two modes of treatment are
to be pursued to suit the kind of product to be obtained.
By the first mode, a fair, yellow sugar, of a quality equal to
that of the ordinary brown sugars of commerce, is the result.
By the second, white sugar, or any grade intervening between it and the crude article, may be obtained.

As a prerequisite to success by either method, the crystallizing and draining rooms should be uniformly heated to
a temperature of not less than 80° F. (26·6° Cent.). To
secure this, a close room is needed, opening by a door into
another apartment instead of by an outside door. The
crystallizing vessels should be ranged along the sides and
a stove placed in the center.

Crystallization and drainage should be performed in the
same vessels, and their form should be such as to conduce
to both these ends.

1. Crude sugar of a good quality and large grain will
uniformly result from well defecated syrup of the proper
density at a temperature of 80° to 90° F. by means of
slow crystallization and natural drainage. The vessels

(111)

should be shallow to admit of the speedy downward passage of the molasses through the crystallized mass, and their bottoms should be inclined sufficiently to secure its rapid transmission to a common outlet. They should be of a uniform size, and in order to secure a large-grained crystallization, should be made moderately large. Vessels conforming to these requirements may be of various forms, but for convenience and general efficiency, I give the preference to a form of vessel which the experience of nearly a century has not modified for the better. I refer to Dutrone's Crystallizing Box, a description of which I give in his own words, as translated by McCulloh (Report, p. 286).*

"Experience has proved to me that the quantity of matter which combines the greatest number of advantages in the crystallization of cane sugar, is fifteen or sixteen cubic feet, for which reason the dimensions given to the crystallizing vessels are five feet in length by three feet in breadth. The bottom is formed of two planes, inclined six inches, the intersection of which form a groove in the middle. In this groove are twelve or fifteen holes of an inch in diameter, to permit the syrup to flow out. The depth is nine inches at the sides and fifteen inches at the center. The vessels should be made of boards one inch thick, and lined with lead" (or better, coated heavily with iron paint). "Before lining it, the holes should be bored in

---

* Précis sur la Canne, page 184. Paris, 1790.

To Dutrone la Couture belongs the honor of originating and introducing many most valuable improvements in the art of sugar making; his work embodies the results of conscientious and laborious research, and although now out of print in its original form, it has been partially reproduced in the English works of G. R. Porter and others. In spite of the crudeness which marked the chemistry of his day, Dutrone has given us a work that yet merits the distinction of being "the most able treatise ever written on the culture and manufacture of sugar."

the groove, and burnt out with a hot iron from the inside, so as to form a small cavity surrounding the hole, in consequence of which not a drop of syrup will remain after draining. . . . Such vessels combine every possible advantage in crystallizing and purging with the requisite strength.

"The crystallizing vessels rest upon strips of wood two inches thick and three inches broad, which are fastened to and supported by upright posts eight or ten inches high, at the distance laterally of ten inches from the middle line. Troughs connecting with a cistern on a lower level receive the molasses as it drips from the sugar."

These vessels, when filled to within 3 inches of the top, will hold about 75 gallons of syrup for granulation, weighing nearly 1000 pounds, of which one-half, or 500 pounds, will be good dry sugar. The depth of the crystallizing mass in these boxes may sometimes be diminished to 3 inches at the sides where the bottom is most elevated, and 9 inches in the center, when there is reason to apprehend any difficulty of drainage by reason of the presence of an undue amount of grape sugar, or otherwise. After the molasses has all drained out, this depth will be much diminished, and the large surface of sugar exposed permits it to dry speedily.

The number of these boxes that will be required will of course depend upon the amount of work to be done, and the length of time that must elapse before they can be refilled and used again. Two weeks is as short a time as can be reckoned upon for the completion of the crystallization and drainage. It will be found that one of these vessels will be required for each 450 or 500 gallons of juice delivered by the mill during that period.

Close the openings in the bottom of the box with long, smooth wooden plugs, abruptly pointed, which may be allowed to project through the holes into the inside of the box two or three inches. Range the boxes in order on the supporting rack, around the sides of the room, and over the dripping troughs, which are so arranged as to convey the molasses into a painted wooden or tin gutter, and thence into a cistern. The dripping troughs may be simply short open conductors of the same materials.

In twenty-four hours after the thick syrup has been passed into the crystallizing box from the cooler, the formation of crystals of small size will generally have commenced. They may then be seen along the edges of the yet liquid mass, but on the bottom of the box they will be found in greatest abundance, and may be detached and brought to the surface at the shallow sides of the box, by means of a knife blade or the wooden scraper, which should be always at hand. The last-named implement is simply a long paddle of ash or hickory wood, with a stout handle and thin blade. With this the fine crystals should be loosened from the bottom and sides, and stirred into the mass, so as to distribute them as equally as possible through it, that they may act as nuclei for the formation of larger crystals. Generally in twenty-four hours after this operation, and often in less time, the crystallization will have pervaded the entire mass. When this is found to be so, then gently withdraw the stoppers and permit the molasses to drain. The sugar will be dry in ten days or less thereafter. It may then be shoveled into boxes or barrels, and the crystallizing boxes refilled.

CHAPTER XVIII.

PROCESS OF MANUFACTURE (CONTINUED).

Drainage—In what it consists—Mode of Working—Extent to which the Production of the Higher Grades of Sugar may profitably be carried in Draining Vessels—Sugar Moulds—Description of the Process of Natural Drainage and the Mode of preparing White Sugar—Reboiling and Crystallization—Summary of Conditions necessary to Success in Liquoring Sugars—Separation of the Molasses by Mechanical Pressure and Other Means—The Centrifugal Drainer.

NATURAL drainage does not remove all the molasses in contact with the crystals of sugar, and hence sugars prepared by that mode are always more or less impure. Molasses is hygroscopic, or attracts moisture from the air; hence the sugar becomes more or less moist and "heavy." The crystals are always in some degree colored, because of the molasses which coats them, and fills the interstices, and for the same reason they have not the pure sweet taste of white sugar.

The purification of sugar is effected by simply washing the crystals with a liquid that is capable of expelling or taking the place of the molasses without dissolving the sugar. This liquid is syrup of greater purity than that which is to be expelled or washed out. It is poured over the top of the mass of impure sugar contained in a vessel which admits of drainage, and is carried down by the force

of gravity through the porous mass, displacing the molasses
as it descends, and finally driving it out at the bottom. By
continuous "liquoring," the term technically applied to
this infiltration, the whole mass of sugar operated upon
may be made perfectly white and chemically pure. But
except in large and well equipped establishments it will be
found neither necessary nor most profitable to carry the art
to this degree of perfection. The aim of the manufacturer
should be the production of a series of graded sugars, corresponding in color and quality to the successive strata
observed in a loaf of sugar which has been subjected to
the liquoring process, to the extent only of rendering the
upper part of it white, while the inferior portions gradually deepen in color from a light yellow to brown at the
tip. The mode of doing this is comparatively simple, and
is precisely the same as that now practiced on the sugar
estates of Cuba and elsewhere. Well crystallized sorghum
sugar requires for its purification no modification of this
mode. The directions given below are adapted to operations of any degree of magnitude, and may be followed
with success by any one who has grown but a patch of
cane, and treated it as already described.

Provide a number of iron sugar moulds that will hold
about 12 gallons each. Smaller moulds may be used, but
the after-treatment of the sugar in these vessels requires
more time than in those of the larger size, and the rapidity
of the crystallization and smallness of the grain of sugars
formed in small moulds is an objection to their use. They
should have received two or three coats of iron paint
and been thoroughly dried and hardened some time before being used. In place of these, conical vessels of a
similar capacity of earthenware or wood may be substituted for experimental purposes, but for ordinary use those
of iron will be found the cheapest and best. For each

mould a pot of earthenware to contain the drippings of the sugar (molasses) must be provided. These pots should hold at least half as much as the moulds.

Range the moulds upon the pots in rows around the walls of the crystallizing-room, and fill them in succession from the cooler. The same method of producing a uniform crystallization as already described, by diffusing the crystals after they have begun to form on the surface and at the sides through the concentrated syrup by means of a long wooden spatula, should be resorted to afterward; but this should not be done oftener than once or twice, or it will defeat its intended object, and cause the production of a mass of minute grains from which the molasses will not readily drain. If from any cause crystallization in minute grains has been induced, the sugar should be kept in a warm place (90° to 100° F.), and natural drainage encouraged as much as possible. When it has ceased, or nearly so, melt and reheat the sugar along with a very small quantity of water to the striking point, and recrystallize under proper conditions.

When the mass in the moulds has become solid, remove the plugs and allow the molasses to drain. If the drainage is sluggish, thrust a large pointed wire or brad awl into the mass from below and withdraw it. This will break up any incrustation which may have formed at the bottom of the mould. After a period of from two to four days, the greater part of the molasses will have dripped out. Then the crust which has formed upon the surface of the sugar, and a portion of the granulated mass below it to the depth of about an inch, are to be scraped off by means of an instrument called a "bottoming trowel," a trowel with a circular blade, leaving the surface of the sugar level or slightly hollowed in the center. These "green cuttings," as they are termed, are put into a pan and kneaded with

cold water to the consistency of a thin paste, in which a
large part of the sugar remains undissolved.   The sugar
in this half liquid form is then to be replaced upon the surface of the mass in the moulds.   Immediately the liquid
part of the mixture, which is a saturated solution of sugar,
begins to penetrate the mass below, displacing the molasses,
and carrying it downward.   At a particular point in the
cone, at which the force of gravity is balanced by the capillary attraction, this downward movement will cease, and
the remainder of the sugar below that point will still be
charged with molasses.   But the syrup derived from the
mixture of water with the "green cuttings" was not pure,
so that the sugar in the upper part of the mould, although
much improved in quality by the application, is not white.
To render it so, and to drive out the remainder of the molasses from the lower part, the process must be repeated,
but this time with a solution of pure sugar.   This pure
liquor could not have been used successfully at the first,
because the molasses to be displaced differed from it too
much in density for displacement to occur.

Prepare a liquor by dissolving so much white sugar in
pure water, that when boiling it will mark 32° Beaumé's
saccharometer, or 36½° Beaumé at the temperature of
60° F. About two quarts of this liquor, cold, should now
be poured upon the surface of the sugar in the moulds
treated as above described.   This liquoring should be
repeated two or three times at the option of the operator,
or according to the purity of the sugar, and at intervals
of from 12 to 24 hours, varying. with the rapidity with
which the liquor penetrates the mass of sugar and disap-
pears from the surface. The drippings of the sugar, after
the second application of the "white liquor," should not
be allowed to commingle with the molasses which has previously drained off, but be carefully collected and set apart

to be diluted to 18° Beaumé, passed through the boneblack
filter, concentrated, and recrystallized.

If the concentrated syrup with which one of these sugar
moulds was originally filled, weighed 140 to 150 pounds,
70 to 80 pounds of sugar will remain, which, treated in the
way described, will consist almost entirely of two grades
of sugar, the upper portion, the base of the cone, will be
white, and the lower part a very light yellow, with a brown
tip.   If a less quantity of the liquor is used than that
recommended, a smaller proportion of white sugar and a
larger proportion of yellow and brown sugar will be the
result.   If, on the other hand, the liquor is applied until
the loaf is *neat*, or the liquor comes through of its original
color, the whole mass of the sugar will be white, but for
reasons already mentioned, it will not generally be desirable
to extend the operation so far.

After the dripping has ceased, and the sugar has been
turned out of the moulds, the different grades of sugar
should be separated by cutting up the loaf with a heavy
knife.   The assorted pieces are then to be dried during a
day or two upon shelves of a room artificially heated to
110°–120° F., and then crushed and packed away in boxes
or barrels with the grade mark on each.

The syrup of drainage, or molasses, may be reconcen-
trated, and if but little glucose originally was contained in
the juice, it will generally yield about one-third of its
weight in the form of crystallized sugar, but the molasses
will be much inferior to the first product, and drainage will
be more difficult.   As long as the market price of syrup
compares as favorably with that of sugar as it does at
present, it will be found most profitable to crystallize but
once, and rework only the drippings of the last liquorings.

It should be remembered that the successive liquorings
are not a source of much loss in well-regulated establish-

ments, since nearly all the sugar in the last drippings is capable of being recrystallized and purified, or reworked to form a new liquor of a somewhat lower grade.   But there is a loss, nevertheless, in the degradation of the white sugar to form a liquor that unites to a greater or less extent with the impurities and coloring matter which it does not wholly displace.   This may be avoided.   A white liquor may be prepared from ripe juice of a known degree of purity, by carefully clarifying it and passing it twice through the filter, if necessary, to remove all coloring matter, and finally concentrating it to 32° Beaumé, boiling hot.   In whatever way prepared, a sufficient quantity of the liquor should be provided beforehand, and kept in a perfectly purified vessel in a cool place.

Iron moulds are to be preferred to those of wood or earthenware.   They are sufficiently strong, little liable to injury, convenient to handle, and easily emptied by exposing them for a few moments to a heat sufficient to liquefy the crystals immediately in contact with their inner surface, and then inverting them.

A summary of the conditions essential to success in the operation of *liquoring* sugar, is thus given by M. Payen, a distinguished sugar manufacturer and chemist.

"1st. That the liquor be sufficiently charged with crystallizable sugar to dissolve little or none in filtering.

"2d. That the density of the liquor be nearly the same or very little less than that of the displaced syrup, for if too dense it will flow badly; too dilute it would escape without removing the syrup or molasses adhering to the crystals.   To attain this condition, the sugars used for preparing the *liquor* must be more impure in proportion as those to be liquored are so likewise ; for saturated syrups are more dense and viscid when they contain uncrystallizable sugar.

"3d. That the crystallization in the moulds be regular and not too compact; and to this end, that it commence and terminate in the same vessel.

"4th. That the temperature of the room in which the liquoring is performed should not vary much, and be at least 70° F."*

The practice of forcing out the molasses by mechanical pressure has commonly been resorted to by sorghum manufacturers, but the necessity of adopting this mode of drainage is an evidence of an inferior crystallization, resulting from unripeness of the cane, imperfect defecation, injury by heat, or other like causes. Natural drainage will not take place with facility in such cases, and the choice of the operator lies between the production of a dark, clammy and inferior sugar by the application of the press, or the melting and reboiling of the refractory mass to form syrup.

In such cases, a better quality of sugar can be made on a small scale by intimately mixing with the plastic mass of sugar and molasses, after pressure has been applied, a small quantity of water, which dissolves the molasses and coloring matter more readily than the crystals of sugar. Pressure is then quickly applied a second time, and the liquid part being thinner and less viscid than before, is more readily driven out. The objections to this mode of drying the sugar are that a single repetition of this process is seldom sufficient to produce a fair sugar from a mass which would not naturally drain, that in the successive washings and pressings a great quantity of the sugar is dissolved, and is rendered comparatively worthless by being mixed with the very impure molasses from which it cannot again be recovered, and that this operation involves a great

---

* Quoted by McCulloh, Report, p. 290, from "Cours de Chimie Elementaire et Industrielle, par M. Payen."

waste of time as well as of materials, and is very injurious to the grain of the sugar. In all cases when the choice must be between drainage by pressure and the reduction of the sugar to the condition of syrup, the former alternative should generally be accepted; but I would recommend that the sugar after the first pressure be washed but once, with a small quantity of cold water, and then pressed a second time. If it be then melted and recrystallized in the proper manner, a fine quality of sugar, of large, sharp grain, will be the result. This sugar will drain well either in moulds or the crystallizing boxes. This plan is not wasteful, for the drippings form a good quality of syrup. There are other means of drainage that might be resorted to more or less successfully, such as driving the molasses through the crystallized mass by atmospheric pressure toward a vacuum formed by the action of an air pump, as is the practice in some refineries. But of all artificial methods of drainage, by far the most rapid and thorough is that which is accomplished by the use of the centrifugal mill.

This simple machine has been adopted very generally of late years in sugar manufacture, both in tropical countries, and in France and Germany. It is one of a number of improvements by means of which, in Cuba and Java, the sugar cane is now made to yield almost twice the quantity of sugar which was obtained from an equal weight of the canes half a century ago. It seems to be admirably adapted to meet the wants of the sugar manufacturer at the North. Although the method by natural drainage is not likely to be supplanted by any other, when it is conducted under favorable circumstances—that is, when a uniform and sufficiently high temperature can be maintained in the draining-room, and a good crystallization has been secured—it is extremely convenient as well as economical for the operator to have at hand a machine by which obstacles to natural drainage, whenever exhibited, may be readily over-come. The unequaled celerity with which it does its work will enable the Northern sugar grower to achieve a result which heretofore could not be secured even in working up the comparatively pure sap of the maple, and which is now attained only in Southern sugar countries, namely, the pro-duction of perfectly dry, crystallized sugar in the evening, from cane which in the morning of the same day was standing in the field !

The essential parts of the centrifugal drainer are a cyl-indrical vessel or metallic drum, with a tight bottom, and closed at the top by a movable cap, the cylinder being fixed in an upright position upon a frame, and within it a smaller cylinder of wire gauze, or sheet metal pierced with small holes, supported upon a vertical shaft, which is made to re-volve at a high speed, the motive power being communi-cated by a belt.

The mass of sugar to be drained must be warmed to about blood heat, and placed within the inner cylinder, which is then made to revolve. The half fluid mass becomes distributed over the inner face of the cylinder, and the ve-locity is rapidly increased. The sugar is retained within the revolving screen, but the molasses is thrown out by cen-trifugal force, and caught in the outer case, from which it is conveyed away by a spout.

In the course of two or three minutes the motion may be checked, and the inner cylinder may then be lifted up, the dry sugar turned out of it, another charge inserted, and the process resumed.

CHAPTER XIX.

SUGAR MILLS.

Composition of Sorghum Cane Juice—Extraction of the Juice—Sugar Mills—Percentage of Juice ordinarily extracted by Mills—Loss from Inefficiency of Mills—Obstacles to the complete Removal of the Juice—To what extent they may be overcome—The Common Three Roll Mills at the South—Mean Results of Experiments with these Mills in Guadaloupe—British West Indies—Cuba, Java, and Louisiana—Importance of further Improvements—Necessity of Grinding at a low rate of Speed—How the Advantages of the Process of Maceration may be successfully combined with those derived from the use of the Mill—Mills with Four Rolls provided with a Pipe for the Injection of Jets of Steam into the Begassa—Value of these Improvements.

IF a ripe stalk of Chinese cane, the juice of which clarified is of the density of 9° Beaumé, be thoroughly dried by artificial means, in air the temperature of which does not exceed 212 Far. (100° Cent.) it will be reduced to about 27 per cent. of its original weight, the remainder—73 per cent., or nearly three-fourths, being water which has passed off by evaporation. In the dried stem there is left the saccharine matter constituting about 14·5 per cent. of its original weight, and 12·5 per cent. of woody fiber-starch, albumen, salts, etc.

The fresh, undried stalk, therefore, contained 87·5 per cent. of its weight of saccharine juice.

The ordinary crushing mills extract but 50 to 60 per cent.

(124)

of the whole weight of the stalk; most of the lighter class
of mills do not average more than 50 per cent. A perfect
machine would produce 87·5 pounds of juice from each 100
pounds of stalk. If in practice but 50 pounds are obtained, 37·5 pounds, or *more than* 43 *per cent. of the whole
amount of juice originally contained in the stalk, is still
retained in the trash.*

We have thus revealed the astonishing fact that about
three-sevenths of the whole product is utterly wasted at the
outset, in consequence of the imperfect means ordinarily
used in extracting the juice. Such a loss is enough to ship-
wreck any industrial pursuit of common magnitude, and it
is a proof of the importance which sugar growing at the
North is destined to assume, and of the firmness of the basis
upon which it rests, that it has proved highly profitable
under such circumstances.

Land which has heretofore yielded the very moderate
amount of 140 gallons of crude syrup per acre, worth in
that condition 60 cents per gallon, or $84, would yield if
the means of extracting the juice were perfect, 200 gallons
per acre, worth $120; and land which, with better cultiva-
tion, has ordinarily· yielded 200 gallons per acre, would
yield 350 gallons.

These facts demand more attention from planters than
they have hitherto received. Here, in the case last mentioned, is a loss of 150 gallons of syrup per acre, the greater
part of which is recklessly, and in many cases, ignorantly
incurred.

It is true that a part of this waste is inevitable. In the
manufacture of sugar from the Southern cane, in which the
most highly improved and powerful crushing mills have
been used, the loss from this source is still very great, and
in many cases much greater than need be, on account of
slovenly practice, yet it is still far less than is commonly

incurred at the North in the sorghum manufacture. Sorghum and the Southern cane are composed of about the same relative proportions of juice and woody fiber; and although the obstacles to the complete removal of the juice are greater in the case of the latter, on account of the joints being more numerous, and the stem generally of closer and more unyielding texture, yet they are the same in kind, and we may adopt here without hesitancy suggestions, which in this instance, experience with cane mills at the South has dictated, and proved to be valuable.

The importance of improving to the highest possible degree the machinery for extracting the juice, forced itself upon the attention of the planters as soon as the composition of the cane became accurately known. Within the last half century nothing has been omitted in the construction of mills of the common form with three rolls, that would in any degree add to their power and efficiency. The results arrived at by careful trial in a great number of instances, with these mills in different countries, have been pretty uniform. Some of the most valuable and trustworthy statements, selected from different sources, are given below.

In the Island of Guadaloupe the results of 44 trials performed with 17 water mills, 15 wind mills, 7 mills with horizontal cylinders, and 5 steam mills, was

61·8 per cent. of juice by hydraulic mills.
61·2 " " " those with horizontal rolls.
60·9 " " " steam mills.
59·3 " " " water and wind mills.
59·2 " " " mills with vertical cylinders.
58·2 " " " mills of animal power.
56·4 " " " wind mills which are still employed
to some extent in Guadaloupe.

59·3 per cent. may be assumed as the yield furnished by

the kind of mills which are the most numerous. The loss amounts to about one-third of the whole quantity of juice contained in the cane.*

In the British West Indies, the loss by reason of the inefficiency of mills, is stated to be at least one-third of the juice. "Of the 90 per cent. of sweet juice which the cane contains, only 50 to 60 per cent. are usually expressed."† In Cuba and Java the yield was no greater, previous to the introduction of certain recent improvements.

In Louisiana, "experiments made upon the plantation of Mr. V. B. Marmillion, 25th November, 1842, with a first-rate horizontal mill, moved by a steam engine of 16-horse power,"‡ gave the following results:

Variety.	Juice. Expressed.	Begassn.	Density of Juice. (Beaumé.)
Ribbon Cane	63·1	36·9	8·25°
Violet Cane	63·0	37·0	8·5°
Gray Cane	64·0	36·0	7·75°
Otaheite Cane	65·8	34·2	8·5
Creole Cane	67·8	32·3	9·0

M. Avequin, who makes the foregoing statement, adds: "I could mention other experiments made for the same purpose in other mills, which experiments have clearly demonstrated that a large number of planters in Louisiana obtain from 63 to 64 per cent. of juice."

The following is an extract from an article written by J. P. Benjamin, Esq., of Louisiana, and published in De Bow's Review for January, 1848:§

"I have taken some pains to ascertain during the present season the yield of juice from our mills of ordinary con-

* Peligot, Report, etc. Paris, 1842.

† Johnston, Chemistry of Common Life, vol. i. p. 209. Appleton, N. Y.

‡ De Bow's Review, vol. vi., No. 1, p. 31.

§ McCulloh's Report, p. 607–8.

struction.   I found the yield from the three-roller mills of
average size and run at a speed of $3\frac{1}{2}$ revolutions per
minute, to be 61 per cent.; whilst from another of very
large size, of which the rollers were $5\frac{1}{2}$ feet in length, and
28 inches in diameter, and which was run at a speed of
only $2\frac{1}{2}$ revolutions per minute, the yield was 66 per cent.,
the begassa being delivered from the latter almost pulverized and apparently dry.   These results are undoubtedly
much more satisfactory than would have been afforded some
years ago; still, they show that after all the care bestowed
in raising our crops, from one-fourth to one-third of our
produce is absolutely lost.   And if we take what I believe
to be a fair average of the yield of juice in sugars, that is,
if we assume that one-tenth of the weight of the juice is
the product in crystallized sugar, we find that we obtain
only $6\frac{1}{2}$ per cent. of the weight of the cane in sugar, whereas
chemical analysis shows that it contains 18 per cent."

Is there then no means by which the yield of juice may
be increased?   Mr. Benjamin's experiment indicates a decided increase by *diminishing the speed of the rolls.*   On
this subject Prof. R. S. McCulloh makes the following valuable observations,* which are as applicable to sorghum
manufacturers at the North, as to those for whose benefit
they were originally intended:

"During the grinding season, when the chief object of
many seems to be rather to hurry through with the crop as
rapidly as possible than to obtain the largest quantity of
sugar of good quality, the grinding of the cane is often
performed in a very improper and ineffectual manner. * *
*   *   *   A general spirit of haste and hurry pervades
the whole force of the plantation, and in consequence much
is done without system.   In grinding, to obtain the proper

---

* McCulloh's Report, p. 204.

yield, the most perfect system should be strictly observed. The mill should not only be regulated to a uniform speed, but should be of sufficient power to work off the entire crop without requiring the speed to exceed the rate which will permit the juice to flow off readily before the begassa is liberated from the mill, so that this begassa may not in expanding act as a sponge, to absorb a large proportion of the expressed juice.   The importance of grinding at a low speed is rendered evident by the following mean results of numerous experiments made by the Marquis de St. Croix, a very intelligent planter of the Island of Martinique, and which I extract from a work* he has written on the manufacture of sugar.

"With the same mill, and its rollers set in the same way, the juice obtained constituted 45 per cent. of the weight of the canes ground when the rollers made six revolutions in a minute, and 70 per cent. when the velocity was only two and a half revolutions per minute; a difference of 25 per cent.

"As the surface developed is for an equal number of revolutions proportional to the diameters of the rollers, M. de St. Croix asserts that a good result will be obtained by rollers which develop a surface of four or five yards in length per minute, so that a roller of two feet diameter should make from two to two and a half revolutions per minute.

"And if it be objected that this velocity is insufficient to express the requisite amount of juice in a given time, *then the length of the rollers should be increased*, and, if necessary, also the power of the engine.   It is the quantity of juice required per hour, under circumstances the most favorable for perfect extraction thereof, which should determine the power of the engine to be employed."

---

* Fabrication Actuelle de Sucre aux Colonies.   Paris, 1843.

It has been proved that by the process of *maceration*, the saccharine matter may be extracted almost entirely from the canes. This process consists in cutting the canes into thin transverse slices, placing them in a succession of boxes with perforated bottoms, one over the other, and pouring into the highest of the series hot water, which as it descends gradually through each removes the greater part of the sugar. But this mode of working is tedious, the hard coating of the canes soon blunts the edges of the circular knives used in slicing, and the cost of evaporation is much increased. Although impracticable for these reasons, it is suggestive of an improvement in the ordinary mode of expression, which has already been put in practice in some sugar countries with very gratifying results. One mode which suggests itself is to steep the begassa in hot water, and press a second time. This is liable, however, to several weighty objections, such as the additional labor and time to be expended in rehandling the crushed cane, and in passing it a second time through the mill, the liability of the very dilute saccharine solution to fermentation and the greatly increased cost of evaporation.

The efficiency of the mill may be increased by increasing the number of the rollers. When mills with five rollers were used, arranged three below and two above, a yield of 70 per cent. of juice was the result, but the very considerable increase of motive power required, prevented them from coming into general use. But mills with four rolls, placed in pairs one directly above the other, with little if any increase of motive power above that which is required for an ordinary three-roller mill, produce 70 to 75 per cent. of juice at one operation. Such mills have been used in Louisiana, yielding the very best results.

If to a mill of this form be added the improvement originally proposed by M. Payen, the distinguished sugar manufacturer and chemist, which is certainly practicable and valuable, it would, I think, leave nothing to be desired in the crushing apparatus. This improvement consists in introducing between the feed and discharge rolls of the mill, or above them, "a pipe placed longitudinally, pierced with numerous small orifices to permit jets of steam to issue, which would be absorbed by the begassa, and acting by displacement, allow the juice to be expressed, which is firmly held by capillary forces in the spongy tissue of the begassa; the steam pipe being supplied from the boiler of the engine, and the mill inclosed in a case of sheet-iron to prevent loss of steam."

The softer texture of sorghum cane warrants the statement that at least 75 per cent. of the weight of the stalk could be extracted by such means, without loss of time or any considerable increase of expense.

It cannot be doubted that a mill so constructed, and possessed of such advantages, would, if brought into general use, add one-fourth to the annual production of sorghum in the United States.

CHAPTER XX.

Chemical Influence of Lime upon Sorghum Juice—Defecation by Tannic Acid and Albumen—Upon what the Peculiar Action of these Substances, so combined, depends—Conditions favorable to the Use of this Method—Another Method of Defecation—The Action of the Chemical Compound used—The Composition described and the Advantages to be derived from its Use.

PURE quick-lime may be used preferably to soda, potash, or other bases, not only to saturate the free acid in sorghum juice, but, in this process, its employment in the mode already prescribed is indispensable for other purposes, no less important. Its action may be briefly stated as follows:

1st. It destroys acidity. Fresh sorghum juice always contains a variable proportion of free acid or acidulous salts, which should always first be saturated or neutralized : the deleterious effects of this acid are : 1. It communicates its peculiar harsh taste, resembling that of unripe fruit, to the syrup. 2. It prevents the separation by heat of the albumen, another substance dissolved in the juice most inimical to the constitution of the sugar, and a most efficient promoter of subsequent fermentation in the syrup. The albumen is held in solution in hot syrup by the acid, but lime, by neutralizing the acid, causes it to coagulate, and it rises and is removed with the scum. 3. This acid has a peculiarly unfavorable influence upon the crystallization of the sugar (independent of its effect in pre-

(132)

venting the coagulation of the albuminous matter) in causing the crystals to be small, ill defined, and difficult of
drainage, and although it is not certain that it causes directly the conversion of cane into uncrystallizable sugar
during evaporation, as some have supposed, it is otherwise sufficiently obnoxious, and lime, by saturating it, renders it comparatively harmless.

2d. Lime in excess converts grape sugar in cane juice
into glucic acid, glucate of lime, and finally molassate of
lime, a peculiar dark-colored soluble substance, uncrystallizable, and without sweetness. In making sugar, this transformation is not to be dreaded, inasmuch as the molassate
of lime in solution is much more liquid than the dissolved
grape sugar from which it was formed, and facilitates
drainage by rendering the molasses less viscid than a solution of grape sugar.

3d. Upon cane sugar the action of an excess of lime is
much less energetic. Four parts of sugar by weight unite
with one of lime, forming sugar lime, or saccharate of lime,
a soluble substance, which becomes coagulable by heat and
passes into the scum. But unless the excess of lime used
be very great, the quantity of sugar lime formed will be
very small, and in a heated solution containing both cane
and grape sugar, the lime does not attack both, but decom-
poses the grape sugar, leaving the cane sugar uninjured.

4th. During evaporation, the decomposition of a small
quantity of neutral salts of ammonia and potash present in
the juice is effected by lime, which has a stronger affinity
for the acids than those bases, the ammonia being given off
in the steam, and the potash remaining in the solution.
The potash, if in considerable quantity, would have a tendency to render the sugar deliquescent, but as the proportion of potash thus liberated is very small, the influence
which it exerts is inappreciable, or reduced to nothing, if, as

Scoffern asserts, it assists in the conversion of a portion of grape sugar into glucate and molassate of potash, which, like the product of the decomposition of grape sugar by lime, is to be preferred to a solution of grape sugar itself, being less viscid. The decomposition of organized sub-stances containing nitrogen also, is inferred to take place during the evaporation.

After complete neutralization of the acid by the use of lime, the juice must be defecated or deprived of a number of substances which, if allowed to remain, would totally prevent the crystallization of the sugar, besides being pro-ductive of other injurious effects. I here recommend two methods for common use, giving the preference to that first mentioned only when its employment is indicated by certain favorable conditions described in Ch. XXX. It effects the separation of a very large amount of impurities, and particu-larly those which are the chief obstacles to crystallization.

DEFECATION BY TANNIC ACID AND ALBUMEN.

The peculiar advantage derived from the use of these substances consists in the removal of all glutinous, albu-minous, mucous, or gummy substances, together with all feculancies and other matters mechanically suspended in the juice, at the same instant, after heat has been applied, and they are firmly incorporated with the scum, which is not again permitted to remain in contact with the clarified juice. In the method of Wray, described in a subsequent page, one of these agents, tannic acid, is used; but in that case it is necessary to add a second dose of lime; the tannic compounds are suspended in the liquid, and are separable only by repeated filtrations, and much time is lost and much trouble incurred before the juice can be prepared for evap-oration. Besides, the clarification is by no means as perfect as when albumen is employed. Defecation is accomplished by albumen and tannin without loss of time. Each of these agents possesses in a pre-eminent degree those qualities in which the other is deficient. The tannic acid unites with and renders insoluble all that most pernicious class of substances, gluten, mucilage, etc., upon which albumen has no chemical action, and the albumen employed in sufficient quantity not only envelops and carries to the surface, when coagulated, all feculancies and mechanical impurities in the liquid, but also all the suspended compounds of tannin, which by other means are so difficult to separate, and forms an insoluble compound separable in the same manner, with any excess of tannin which would otherwise remain dissolved.

DEFECATION BY ALBUMEN, ALUMINA, AND TANNIN.

This mode consists in the addition, to the cold juice, of alumina, united with a definite proportion of albumen and tannin. The tannin as well as the albumen dissolve readily in the juice, which should previously have been carefully neutralized. A preparation composed of a combination of these substances in the dry state, in the proper proportions, is most convenient. Its peculiar action upon the juice immediately follows upon the application of heat. A reaction takes place, and a new group of substances is instantly formed, all of which, upon the application of heat to the liquid, become insoluble. The alumina has a most powerful attraction for coloring matters, which it takes up and unites to itself in the insoluble form. Its use as a decolorizer has long been known. Formed from alum, its use as a mordant, in fixing colors, is as old as the art of dyeing, and it has been applied to a very limited extent also as a defecator in sugar manufacture from the

cane of the tropics. But its employment on the plantations was attended with difficulty, inasmuch as it was
deemed necessary to heat the juice to the boiling point in
a deep tank, and then to allow it a long period of rest before the precipitation of the alumina and the impurities
with which it was combined could fully take place. The
loss of time in the heating of the juice, and in the slow
precipitation, together with the difficulty of separating
from the bulky sediment the portion of juice in immediate
contact with it by any means that could conveniently be
applied, were obstacles which were generally thought to
outweigh the advantages derived from its use.

While alumina is one of the most energetic decolor-
izers known, not excepting even animal charcoal, it does
not remove from cane juice by ordinary precipitation some
impurities of another kind. These are feculancies very
abundant in sorghum juice for which the alumina possesses
little or no attraction—and which do not naturally separate
by heat—glutinous and gummy matter are also present in
considerable quantities. And if these are permitted to
remain until the syrup becomes more dense by evaporation,
they cannot afterward be entirely separated, and they impair the quality of the syrup not only by reason of the
viscidity which they impart to it, and by the destruction
of a portion of the sugar through the chemical changes
which they produce when assisted by heat, but also by the
precipitation upon the bottom of the pans of a peculiar
gummy sediment which, adhering to them, forms, with earthy
matters, the hard scale so common where the defecation
has been imperfect. This differs from ordinary lime scale
not only in its composition, but also in its liability to burn
at once upon the bottom of the pan.

For the removal of this class of impurities, at this stage
of the evaportion, no means that I have employed can be

compared to albumen, assisted by a very small but definite proportion of tannic acid. Albumen, in small quantity, is always
present in sorghum juice, and by its coagulability at a temperature below the boiling point it assists considerably in
its clarification, and is suggestive of the effects which it is
capable of producing when added in sufficient quantity.
But the quantity naturally incorporated with the other
impurities above mentioned is too small to effect the removal of but a part of them, and this effect even is produced
only when the juice has been previously neutralized, for it
is not coagulable in juice containing an excess of acid.

If we compare the results of defecation by means of
alumina, albumen, and tannin separately, we find that each
has a peculiar attraction for certain substances, and that
neither used singly nor in any way combined in pairs, are
they capable of accomplishing the work of the three combined in the proper proportions.   They unite together,
and with the substances that they are designed to remove
from the juice, in the insoluble form, and all are carried
away.   Separately they would act in a different way.   The
alumina being an earthy substance, when used alone, carries
down the coloring matters with which it unites and de-
posits them as a sediment in a liquid at rest.   The tannic
and albuminous compounds, on the contrary, rise to the
surface of the heated juice.   Yet I find that when suitably
combined, all three and the associated impurities rise to
the surface—the superior buoyancy of the albumen, especially, overcoming the tendency of the alumina to sink and
bringing it up along with it.   Each minute particle of the
alumina is a center of attraction for the coloring matter
in the fluid around it, which it seizes upon and forcibly
retains.   When the juice is admitted into the evaporating
pan, the tannin and albumen are dissolved, and the alumina
is suspended in the liquid.   The juice becomes almost im-

mediately heated to the boiling point, and then the albumen and the tannin become visible in the compounds which they form—the former weaving through the liquid mass myriads of gossamer-like filaments, enveloping all the suspended substances in its folds, in which, as in a drag net, it sweeps them up to the surface, and immediately in the form of scum they are carried over into the receptacles provided at the end and sides of the pan—the juice, meanwhile, transparent as glass, being borne onward in the opposite direction. Thus, at the same instant the natural impurities and all the associated defecating substances that, a moment before, were commingled in the juice, are separated from it in the insoluble form.

This compound seems to conform to all the conditions required of a defecator of the juice of sorghum. It may be afforded at a low cost—it is entirely free from any properties injurious to health if by accident, or mismanagement, any of it should remain in the syrup—and it is especially adapted to a system of rapid heating in a shallow pan, so constructed as to remove the scum from all subsequent contact with the syrup, as fast as formed, by a superficial current flowing in a direction opposite to that taken by the clarified juice. It thus, with far better results, renders entirely needless any precipitating tanks, and the loss of time and labor attendant upon their use.*

Pure neutral sulphate of alumina, followed by lime, may be used to decolorize the juice, with advantage, but time must be given for the precipitate to settle, which it does very slowly,—and in point of general efficiency it is very

---

* In order to supply this substance to those who may need it, of the requisite purity, and accurately compounded, arrangements have been made for its manufacture, at a price which will place it within common reach.

far inferior to the defecating compound above mentioned. Alum has been used in the beet sugar manufacture, and has sometimes been highly commended, but its employment should be entirely discarded. It leaves in the juice a solution of sulphate of potash, a salt which, in addition to imparting a bitter and offensive taste, also gives the property of deliquescence to the sugar.

CHAPTER XXI.

ANIMAL CARBON, OR BONEBLACK.

Filtration—Animal Charcoal or Boneblack—Mode of preparing it, and the Condition in which it is employed in this Process—Properties of the Charcoal in this form—Mode of using it—Its Action—Use in removing the Extractive Matter and Harsh Flavor—Decolorizes—Promotes Crystallization—Removes any Excess of Lime, Tannin, etc., etc.—Separates all Soluble Substances mechanically suspended in the Liquid—Mode of restoring the Powers of the Charcoal when exhausted—Different Modes of Reburning—New Method of Revivification—Substitutes for it in Filtration.

AFTER defecation by the means indicated in the foregoing chapter, the cane juice is beautifully clear, and apparently free from all impurities. It is not entirely pure, however, for there still remains a little of a harsh vegetable flavor due to extractive matter, which is separable neither by the process of defecation, nor by the action of heat. The entire removal of this is the primary object sought in the peculiar filtration to which the syrup is now to be subjected. This done, it is evident that a syrup so prepared is a refined syrup of the first quality. It should be even superior to the best refined syrups of the market, because a long and tedious series of operations, including dilution with water, clarification, filtration and reboiling, as practiced by the refiner upon the crude article, is not only made unnecessary, but also the waste, and the injurious effects resulting thereby to both syrup and sugar, are avoided.

(140)

The filter used in this proces differs from the boneblack filters of the ordinary form in its shallowness, the comparatively small quantity of the charcoal used, the fineness of division of its particles, and in some of its chemical properties.

Ordinary boneblack, as used by the sugar refiners, is prepared by subjecting clean beef-bones in iron retorts, or covered pots in a furnace, to a red heat for some hours. An imperfect combustion of the bones takes place, access of air is prevented, all volatile substances which are produced are driven off, and a dull black substance remains behind retaining the original form of the bones, consisting of phosphate of lime associated with a little carbonate of lime, sulphuret or oxide of iron, and silicated carburet of iron 90 parts, carbon 10 parts. The charred bones are excluded from the air until cool, and then are broken up in a mill into small fragments, averaging 0·1 to 0·3 of an inch in diameter, and the finer particles are sifted out. This charcoal is exceedingly porous, and its superior decolorizing power, as well as its chemical action, seems to depend upon the very great extent of surface presented by the carbon to the liquid with which it comes in contact. For the same reason the more finely divided its particles are the more effective it becomes. It is far more energetic in its action than most other forms of charcoal, besides possessing other properties peculiar to itself. All these properties, however, may be variously modified, by the mode in which it is prepared, and its particular form, to suit certain ends.

The action of the filtering material used in this process embraces the following particulars:

1. It removes the extractive matter which imparts to sorghum syrup its peculiar harsh flavor, and which resists the action of heat and all other means that have been used to effect its separation.

2. It decolorizes. Although this, as has already been said, is not a primary object, the color of the previously clarified syrup being very fine, it is still further improved, and the syrup acquires a glassy brightness and purity of color not otherwise attainable.

3. It promotes crystallization in a high degree, and increases the quantity and improves the quality of the sugar. The ordinary boneblack is said to increase the crystallizing property of the juice of the tropical cane in a remarkable degree, in some instances 18 to 20 per cent. Upon the juice of sorghum its influence, in this respect, is very decided.

4. It removes any small excess of lime which may have been added in saturating the free acid in the juice. This property of boneblack is practically of much value. To insure a perfect neutralization of the acid, and the consequent coagulation of the albumen, the juice should be made to exhibit a faint alkaline reaction — or, in other words, a slight excess of lime should have been added. This the filter wholly removes. So also it removes any excess of tannin or alumina that may inadvertently have been introduced.

5. Independent of the above-mentioned properties, it acts as a strainer of the most effective kind for arresting any insoluble substances mechanically suspended in the liquid, such as compounds of lime and fine particles of sand, more or less of which are found in all unrefined syrups.

As the intensity of the action of the carbon may be modified indefinitely according to the state of division to which it is reduced, the density and temperature of the syrup, and the nature and quantity of the impurities which the syrup contains, the boneblack must conform, when used as above, to a certain standard in quality and properties, and it must be used in the manner already described.

The boneblack should be reduced to fragments of considerably less than half the size of those of refiner's boneblack, varying between 0·05 and 0·1 of an inch. The depth of the stratum in the filter drawer need not exceed 6 to 8 inches; the density of the syrup should not exceed 15° to 18° Beaumé's saccharometer, and it should always be passed through hot. The rapidity and perfection of the filtration is secured by the position of the filter immediately over a boiling bed of syrup, where it is enveloped in an atmosphere of steam, by which contact with currents of cold air is prevented, and it is kept continually at a uniform degree of heat, receiving the boiling syrup from the pan above.

This system of refining simultaneously with the evaporation is advantageous in several important particulars. It saves time; the syrup is passed directly from one compartment of the evaporator to the other through the filter. It effects more easily the separation of the impurities than if they had been suffered to remain and form new compounds at higher temperatures. The sugar also is then more readily crystallizable, is of better quality, and may be produced in larger quantity than after reboiling at any subsequent time.

The bone-charcoal should be prepared from large and solid beef-bones, carefully burned, according to the usual method, as already described. It should be neither of an ashy-gray color, nor lustrous black as if glazed, but of a deep dull black, and destitute of both odor and taste. When it has been exposed to too high a heat, it becomes glossy and is partially fused, and when allowed too free access to the air while burning, or while being cooled, much of the carbon is consumed, and in either case it is comparatively inefficient. When not exposed to the proper heat, for a sufficient length of time, a part of the substance of the

bones is not charred at all, and it proves ruinous to the
syrup by communicating to it the peculiarly offensive flavor
and odor of the animal oil and empyreumatic matter,
which has not been driven off. This flavor and odor cannot
be afterward removed from the syrup by any means known.
In burning boneblack a low red heat, prolonged for about six
hours, in the ordinary covered iron pots, is productive of
the best results, and no difficulty whatever need be appre-
hended when the covers of the pots are luted with clay, or
when pairs of pots are used fitting closely together, mouth
to mouth, or when the bones are calcined in retorts.

The powers of the boneblack become exhausted by use,
and as has been mentioned, it is necessary to replenish the
filter daily. But little loss of the material is thus incurred,
however, for its energy may be restored by reburning, and
this process of successive exhaustion and revivification may
be repeated indefinitely, if the very fine particles or dust,
amounting to but a small percentage of the whole mass, be
sifted out each time after reburning, and enough of fresh
carbon be added to supply its place. In order that its
action may not be impaired, the boneblack should always
be washed free from sugar before reburning it. (See Ch.
XVI.)

The revivification of the carbon is accomplished econom-
ically on the large scale in different ways. It may be exposed to a red heat in cast-iron retorts of a conical shape,
placed vertically over openings in a bed-plate in a furnace.
Through these openings the boneblack is discharged, when
red hot, by drawing a slide, and through similar apertures
in the tops of the retorts, they are refilled. Or it may be
placed in a cylinder, and steam, heated to 700° or 750° F.,
driven through it. It may also be reburnt in the iron pots
above mentioned, with success. Large retorts are sometimes made of fire-bricks, of the shape of a square hollow

pillar, or right rectangular prism, closed at the top by a movable iron plate, and below by a double trap door, opening downward. Surrounding this retort is a furnace of brick work, of the same height as the retort, but wider below than above. (Pereira.)

The revivification of the boneblack used in this process may perhaps most readily be accomplished by inclosing it in a cast-iron cylinder, placed horizontally in the floor of the evaporator furnace itself, just beyond the grate bars. The cylinder may be revolved, so as to expose its contents equally to the heat, by means of projections at its ends, outside of the furnace walls, where also there are openings through which to fill and empty the retort.

A method of restoring, or even of augmenting (as is asserted), the powers of animal charcoal has recently been discovered by Mr. Edward Beanes, of England, who introduced some improvements in sugar manufacture in the Island of Cuba. It consists* in treating the exhausted charcoal, when dry and hot, with dry hydrochloric acid gas, which it absorbs with astonishing avidity. Another portion of charcoal is then added to that which has received the acid gas, the combined gas remaining in the pores of the latter is taken up by the former, and the whole becomes neutral: chloride of calcium is formed, which is easily washed out, and the charcoal is then reburned in the usual way. The advantages of this method are said to consist chiefly in the removal of any lime that the charcoal may have taken up without attacking the phosphate, and in augmenting the decolorizing powers of the coal more than 100 per cent. The efficiency of this means has not yet been tested in this country.

Freshly burned charcoal should be carefully excluded

---

* Medlock, in London Chem. News—Sci. American, vol. xiii. p. 224.

from the air until it has cooled, or it may be quenched while hot by sprinkling it with water, taking care to add no more than is necessary for that purpose.

Thorough defecation, according to the method already prescribed, is a valuable auxiliary to the action of this filter, and these operations may be regarded as inseparable and indispensable in this process.

Although many other forms of carbon might be proposed as substitutes for this preparation of boneblack, none of them have been found at all equal to it. for attaining the desired results

CHAPTER XXII.

THE ACTION OF HEAT.

THE application of heat to cane juice until its temperature rises to the boiling point, is capable, unassisted by other agents, of effecting an imperfect clarification. This it does chiefly by effecting the coagulation of that portion of the albumen of the juice not held in solution by the acid, and by the separation along with this albumen of a great part of the leaf green (chlorophyll) and other substances mechanically suspended in the liquid.

During the subsequent evaporation the temperature constantly rises, and when it reaches a certain point the heat causes the decomposition of some of the nitrogenous impurities not before separated. This action of the heat, however, may be very much assisted by the chemical action of lime. At a much higher temperature the sugar itself is decomposed (burnt), and the application of a very strong

(147)

heat at the closing stage of the evaporation is capable of
producing this disastrous result when the great mass of the
syrup is of a much lower temperature than that at which
caramelization of the sugar occurs. The cause of this is
to be found in the imperfect convection of the heat through
the sluggish mass of half liquid syrup. It is not conveyed
away as fast as it is received by that portion of the syrup
which is directly in contact with the bottom of the pan.

Hence the necessity of placing the finishing pan at that
extremity of the evaporating range which is farthest removed from the furnace. The pan, too, should be shallow,
in order that the evaporation may be rapid, and that the
transmission of the heat through it may be as rapid as its
absorption by the lower stratum of the liquid. It should
be provided with a thermometer capable of indicating at
any moment to the eye the temperature, and thence the
density, and the pan should be so adjusted as readily to
admit of the instant removal of its contents from exposure
to the heat, and, with equal facility, of being emptied, replaced and refilled. Such a finishing pan should secure
the best results attainable by a system of open-air evaporation, and the form which most fully meets those requirements in practice, and seems best adapted to common use,
is that of the tilt pan. (D. See *engraving.*) It is merely
a modification of the old French *bascule*, a vessel long in
use among sugar manufacturers in Mauritius, Guadaloupe,
and elsewhere.

A *prolonged* heat is exceedingly injurious to a solution
of sugar. The control of heat over the chemical affinities
which determine the form and sensible properties of most
substances, and especially of organic bodies which are
isomeric, is well known. Sorghum juice contains a number
of substances which are peculiarly unstable in their nature.

Some of these are starch, dextrine (cane gum), grape sugar, and cane sugar. They seem to be mutually convertible in the growing plant, and by the action of heat upon the juice, assisted by merely the *presence* of an acid, are capable of being transformed, the one taking the form of the other in the order above named,—with the exception of the conversion of grape into cane sugar, which no art can accomplish, although such a change takes place with facility in the growing plant. The occurrence of degrading transformations, or such as follow in the reverse order of the substances named, such as that whereby cane sugar is made to take the form of grape sugar, or other inferior forms during evaporation, is common; indeed it may always be inferred to take place when the acid and impurities natural to the cane juice have not been neutralized or removed, although such change may not be immediately perceptible, either to the eye or to the sense of taste. It is a great error to suppose that no such change is taking place unless the effects of the decomposing power of heat are *visible to the eye.* Heat, with the presence of an acid, will convert cane into uncrystallizable sugar, without any marked indications that such a destructive process is in progress.

But a moderate heat alone, if prolonged, is capable of decomposing a pure solution of cane sugar. The experiments of Soubeiran proved this conclusively. He dissolved a given quantity of sugar in a given quantity of water, and applied heat to the solution for 36 hours. "The apparatus was so constructed, that the water given off by evaporation was continually returned to the original solution, by which contrivance the latter was always composed of the same quantity of sugar, or its derivatives, and the same quantity of water, as when the experiment commenced. Gradually

the solution acquired darkness of color, and at the end of
36 hours it had become black. The effects of long application of the heat were: 1. The disappearance of cane
sugar. 2. The appearance of grape sugar or glucose. 3.
The production of carbonaceous powder and acids."

It is impossible, indeed, to prevent the formation of a
small portion of darker colored liquid sugar (molasses) in
a solution of white sugar, chemically pure, which has been
heated to boiling, and afterward recrystallized.

Claiming for the tilt pan no other merits than those which
have recommended it in plantation use for the last hundred
years, I here give place to some remarks by Prof. McCulloh, founded upon the experience of M. Payen, the eminent French chemist, in regard to open-air boiling and the
employment of an evaporating vessel somewhat similar.

"It seems to be a disputed point whether or not saccharine solutions and juices are injured in evaporation by
exposure to the air. M. Payen, whose opinion and experience are entitled to great credit, remarks that it is the
more important to refute the belief that syrups are injured
and rendered dark colored in evaporation by an elevated
temperature, or by the *action* of *the* air aided by heat, because these views have been sustained by eminent scientific
men, and given rise to most ruinous speculations. In
corroboration of his opinion that injury is done rather by
long duration of heat, M. Payen adduces the facts that
boiling for thirty or forty-five minutes, according to the old
system, deepens the color, and renders a much larger quantity
of sugar uncrystallizable than rapid concentration in six
or eight minutes by means of a *tilt* or *bascule pan;* that
slow evaporation, by steam, of large quantities of beet
juice *at a temperature below that of boiling water*, far
from producing a better result, gives very dark and per-

fectly uncrystallizable syrups; that slow evaporation *either by an open fire or by a water-bath*, gives equally bad results. As for exposure to the action of the air during concentration, M. Payen remarks that, far from considering it very prejudicial, the effect should be regarded as almost nothing; for comparative experiments made in vacuo, in carbonic acid gas, in nitrogen and in atmospheric air, gave him like results for like temperatures and times of evaporation.

"Again, it is considered as a fact fully established by the use of apparatus, similar to that described above in the manufacture of beet sugar, that saccharine juices do not sustain appreciable injury in concentration by *exposure to the air;* and for this we have the authority of the most intelligent and experienced manufacturers and chemists."*

The truth of the foregoing observations has been fully confirmed by six years' experience in the manufacture of sorghum syrup in the United States. Up to a certain point of concentration, no prejudicial influence whatever is occasioned by exposure to the air or to rapid boiling. But when a temperature of 220° F. is reached, the danger of decomposition of the sugar becomes very much increased, on account of the greater density of the syrup. Nor can this difficulty be fully obviated by any of the usual methods of evaporation at the ordinary pressure of the atmosphere, — for if the evaporation be conducted at a diminished temperature, as by the use of steam at 212° F., or in a vessel immersed in a water-bath—all the evil consequences of the prolonged action of heat will result, and the sugar will be much more injured than by the action of a strong heat at the close of the operation.

---

* McCulloh's Report, Senate Doc. 50, p. 238.

Happily, the problem of rapid evaporation at a low temperature has already been solved.   The application of the principle, that rapid boiling and evaporation will take place at a low temperature under diminished atmospheric pressure, was a great stride in the progress of sugar manufacture.   And now the vacuum pan of Howard, or of others who have variously modified it. is used wherever skill and large capital are combined in the production of sugar from the tropical cane.   In the manufacture of sugar from sorghum, wherever a steam engine is used to propel the crushing mill, a vacuum *finishing* pan of the simplest form would unquestionably be better than any other : the steam furnishing the heat to the jacket of the boiler or to the coils within it, and the engine driving the air pump by which the exhaustion is maintained.   By this means the syrup, previously boiled to the proper density in the open air, is reduced with wonderful rapidity to the proper degree of concentration, at a temperature not exceeding 170° F., the sugar which results after complete crystallization being not only somewhat more abundant, but also superior in fairness of color and sharpness of grain to that obtained by the use of an open-air finishing pan.   At the same time the molasses is reduced in quantity, and what is more important, it is of so much better quality as to furnish, after reboiling by the same means, a second crop of crystals but little inferior to the first.

These advantages will no doubt secure the adoption of the vacuum finishing pan in all sugar works wherein capital is largely invested, but its expensiveness will preclude its general use.   When a vacuum pan is employed, the syrup should pass through the same process in the evaporating range as before mentioned, only it should not be retained so long in the last two pans of the series, but should pass from the tilt pan to the reservoir supplying the vacuum pan when the boiling heat rises to about 220° F.

In large operations where it is desirable to increase the evaporative capacity of the range without unduly increasing its length, it may be divided conveniently into two, placed upon separate parallel flues opening into the same smoke stack, the pan comprising the first and second divisions, A and B being placed upon furnace walls sufficiently elevated above those supporting the pans C and D, to permit the syrup to flow into the head of C from the filtering box. The length of each half of the divided range may be increased to at least 24 feet, by increasing the length of all the divisions except A.

The expensive vacuum finishing pan, although the best for use in large sugar factories, is not equally well adapted to the sphere of operations within which sorghum sugar production will ordinarily be conducted. For working up a crop of less than 75 acres of cane, undertaken by ordinary farm hands, no better apparatus is needed than the double evaporating range above described. Upon it a golden syrup, unexcelled either in color or flavor by the best products of the refineries, may be made with but little expenditure of money, time, or labor, and as much sugar as will be sufficient to satisfy any reasonable expectation, equal in all respects, either for sale or for domestic purposes, to the article which during the past fifty years has been most in demand for common use. If the intelligent farmer, at a season of comparative leisure, expends in the production of a given amount of sugar proportionally more time and care than the large manufacturer, he needs to spend but little money in the purchase and repair of costly apparatus, which can be used only a part of the year. Within his limited sphere of operations, growing and manufacturing fair yellow sugar upon his own farm, and utilizing all waste products, he can successfully compete with the large manufacturer.

Waste of fuel is one of the evils attendant upon the system of evaporation in short pans heretofore much used. In districts where wood and coal are scarce, the cost of fuel is often more than 50 per cent. of the current expenses. Experiment has proved that the expense of fuel, to supply the long evaporating ranges here recommended, is reduced to one-half of that which is ordinarily incurred.

CHAPTER XXIII.

OTHER METHODS.

Description of various Methods of Manufacture heretofore used
or recommended—The Method of Wray described—Its Merits
and Disadvantages—Its Impracticability—Melsen's Method—
Specifications—Its Advantages dependent upon the presumption
that Sorghum and Tropical Cane Juice are identical, which is
untrue—Further Reasons why Bisulphite of Lime, as a Defecator of Sorghum Juice, is of no Peculiar Value.

HAVING in the preceding chapters given the details of a
system of sugar manufacture from sorghum, which has been
uniformly successful in my own hands, and which I have
reason to believe will prove equally so in common use, and
in which all the main practical difficulties that have hitherto
stood in the way have been overcome, I shall now trace out
briefly the outlines of the different processes that have heretofore been used, and define what I conceive to be their
defects, judging from experiment, and the causes of their
failure.

The complex nature of sorghum cane juice, and the
marked peculiarities which distinguish it from the juices
of the tropical cane and the sugar beet should not be
lost sight of in an investigation of the comparative merits
of different methods of manufacture. These characteristic properties require at least a modification of the
modes of treatment usually applied to the Southern cane
and the beet, to the extent to which these properties are

uniformly found to be exhibited. In all cases, the separation of the sugar in the crystalline form from the other constituents of the juice, includes several distinct operations, resolved into an apparently simple routine by practice,—and is in fact throughout a process of analysis on an extended scale, involving in its different stages:

1st. The separation of the crude saccharine solution from the woody and cellular tissues of the plant. This is best accomplished by the mill.

2d. The separation of various impurities from the saccharine solution, and the neutralization or transformation of others of a mischievous nature which cannot be separated. This is the process of defecation.

3d. The separation of the greater part of the water in which the sugar is dissolved. Evaporation accomplishes this object.

4th. The separation of crystallized sugar, more or less colored and impure, from the resulting semiliquid mass or mother-liquor which still contains crystallizable sugar and molasses. This includes crystallization and drainage.

5th. The separation of all remaining impurities from the more or less colored sugar of the last operation. This is commonly the work of the refiner. Chemically pure white sugar is its best result. It involves various processes, the most important of which is that technically known as "liquoring." Few manufacturers, however, will find it profitable to carry the process through this last stage.

As complete defecation or the separation of the saccharine matter from the different substances with which it is combined in solution is the chief difficulty to be surmounted, it will be sufficient at present to call attention to this one point, as I am satisfied that if this be perfectly attained, the different steps which precede and follow it will present no difficulty, as they depend chiefly upon me-

chanical means which are of wider application, and more
generally understood. Defecation requires the use of such
chemical agents, and mechanical appliances, as will cause
the separation of the impurities from the juice without impairing the quality of the sugar—a substance which in
such association is remarkably unstable, and liable to be
either injured or totally destroyed. It should be mentioned also that there are many beautiful methods known
to the chemist, and extremely useful and successful in the
laboratory, which, on account of the expense, complexity
in the use of the means employed, or for other reasons, are
totally inapplicable on the larger scale in practice. With
these at present we have nothing to do.

WRAY'S METHOD

first deserves attention, as it was originally brought into
notice in connection with the earliest attempt to make
sugar from sorghum in this country. At the first view the
merits of this process are obvious. Throughout, it is
simple and easily understood. It consists in treating
sorghum juice — previously neutralized by lime — with a
solution of nut-galls (tannin), which has the property of
precipitating in an insoluble form from the juice of ripe
cane, nearly all the impurities which hinder the crystallization and drainage of the sugar. Tannin, or tannic acid,
the active agent used, is an abundant natural product, and,
as has been already mentioned, may be obtained in a sufficiently pure and inexpensive form for ordinary use. The
defecation is begun and finished in the cold juice, and the
detrimental effects of successive reheatings are entirely
avoided, and the evaporation may be carried on expedi-
tiously to the close without any interruption. These are
important considerations. I will add to them by saying,

14

that the defecation is more complete than can be effected
by the use of any single substance of which I have any
knowledge that may be legitimately employed in common
use for this purpose. As a result of this, crystallization
takes place with great facility. In my hands, samples of
a very light-colored sugar have crystallized in a solid mass
with but little molasses almost as soon as cold.

Yet, with all these advantages, the cause of the failure
of this method, as used by its inventor and others, must, I
think, be apparent to every one who has subjected it to a
rigorous practical test. The tediousness of this process
is the first prominent objection that presents itself. Three
filtrations of the whole volume of cold juice are necessary.
One to remove the precipitate after neutralization by lime;
another to remove the insoluble matter precipitated after
treating with the solution of tannin; and a third, after
again treating the juice with lime in such quantity as to
form a precipitate with all the tannin remaining in the solution. The loss of time consequent upon this series of filtrations, will appear when this process is compared with
any one by which defecation is attained during the brief
period in which the juice passes through the first stage of
concentration, in which no filtration through fine cloth or
some medium of·equally close texture, such as is necessary
in this, is required.

Tannic acid, as is well known, possesses the property of
imparting an inky color to solutions containing any of the
salts of the sesquioxide of iron even in very minute quantities, and inasmuch as contact of the acid juice with iron
implements, such as the crushing mill, etc., would be sufficient to form in it a trace of iron in solution, it has been
objected to this method that a discoloration of the juice
invariably ensues after the addition of the tannic acid. It
will be found, however, that in all ordinary cases, little or

no change of color will be produced in this way; but if there should be, it will be but temporary, and of no consequence. The dark coloring matter being a precipitate which is merely mechanically suspended, and not dissolved in the liquid, is thrown up with the other impurities when heat is applied, and it is removed with the scum.

A much more serious disadvantage is the extreme degree of care that is necessary — much more than would ordinarily be exercised — in apportioning the chemical agents used to the exact effect designed to be produced. Three times — once before each filtration — the juice must be treated with lime or tannin; these must be in due proportion to each other, or to the variable amount of impurities in the juice with which they combine. It is true that the bad consequences of a failure to exercise proper care in the use of defecating agents is not peculiar to this process, but belongs to all others in which such substances are used. Yet, here, the lapse of an extended interval between each application imposes a tax upon the memory of the operator whose attention is divided between this and other matters, and the liability to commit errors through forgetfulness is in this case much greater than in those in which the required additions are made in immediate succession.

The practical difficulties attendant upon the successive filtrations of the cold juice are the chief impediments known to the general adoption of this method, and we are compelled to abandon it with regret, since in other respects its results are very satisfactory.

MELSEN'S METHOD,

which has acquired considerable celebrity in Europe as applied in the beet-sugar manufacture, and also in Louisiana, where it has been used lately to some extent on the plantations, is asserted to possess peculiar merits,—stated by its discoverer in the following words: Bisulphite of lime (the agent employed) acts—

1. As a powerful antiseptic, preventing the production or formation of fermenting matter.

2. As from its affinity for oxygen capable of preventing the changes which the presence of that agent causes in the juice.

3. As an agent which at 100° Centigrade defecates the juice, and removes from it all the albumen and coagulated matter.

4. As carrying away the pre-existing discoloration.

5. As an agent capable in the highest degree of preventing the formation of coloring matters.

6. As capable of neutralizing all the hurtful acids which exist or may be formed in the juice, substituting for them an acid almost inert (sulphurous acid).—(*Extract from Mons. Melsen's Memoir*. Translated by F. G. Clemsen, and published in the Agricultural Report of the Patent Office, 1849–50, p. 404.)

The value of bisulphite of lime in preventing fermentation can scarcely be overestimated; and one great hope of the discoverer of this process, and that which led him to the selection of this agent was, "that in the equatorial regions at least, sugar might be extracted *by the heat of the sun alone.*" The antiseptic power of the bisulphite, preventing fermentation for an indefinite length of time, was thus most prominent in his view; its defecating properties, although important, seem to have been in his view, as they are in reality, entirely subsidiary to that end. As a defecator of the juice of the tropical cane and the beet, its action is not complete, as asserted by himself, as, after the clarification, "there remains a matter which is colored first violet, and afterward brown, by the air or the influence of an alkali." Specifications 4, 5, and 6 (see above), assert facts which are practically of small importance, for by the use of suitable evaporating apparatus after complete clarification no injurious discoloration is subsequently incurred; and by the use of common lime, the hurtful acids may be effectually neutralized.

The sulphurous taste said to be imparted to sugar by this mode of treatment, is not a serious obstacle to its adoption, as it is removed by exposure to the air or by the operation of "claying." In a hot climate, where fermentation is rapid, and the saccharine juice is comparatively pure, demanding but a low defecating power, this method is successful; but its merits are not apparent in this climate in its application to sorghum cane.

The advocates of the adoption of this method by the sorghum growers of the North, proceed upon the assumption that "the constituents of the sorghum and Louisiana cane are very nearly, if not altogether, identical;" and hence it is predicted that Mons. Melsen's discovery "is destined to exert an influence upon the production of sugar from the sorghum no less marked and decided than that which it has already accomplished in the manufacture from the Southern cane." This assumption, it is scarcely necessary to say, is entirely erroneous; the constituents of sorghum and Louisiana cane are by no means identical, and any conclusions based upon such identity, and not verified by experiment, are valueless.

As a defecator of sorghum juice, I have found the bisulphite of lime much inferior to other agents that may be employed. One of the most experienced sorghum growers in this country asserts that he is "unwilling to recommend its use from any evidence yet adduced." (*Ag. Rep.* for 1861, p. 303.) Though inadequate to accomplish the

14*

main object in view, this agent has, I believe, an important use in being readily applicable in preventing fermentation, under circumstances when the immediate application of heat is impracticable.   It is said also by some to have the power of restoring the original taste and color to the juice of soured canes, but the statement "needs confirmation."

CHAPTER XXIV.

THE COMMON METHOD BY HEAT.

Extent to which Defecation takes place by this Means without the
use of Other Agents—The Advocacy of Erroneous Opinions concerning the Action of Heat productive of much Mischief—Sugar
cannot ordinarily be made by such Means.

SEVERAL modifications of a single method of manufacture have been applied for the production of sorghum
syrup; the merits of these are made to depend solely upon
different forms of evaporating surface in order to secure
certain real or fancied advantages from the application of
heat. The principle of rapid evaporation in shallow pans
is certainly correct, but rapid evaporation and the separation of such impurities only as heat can remove, are but a
part of the means requisite to success in the manufacture
of merely a fine syrup, not to mention sugar. If boiling
and skimming alone could do the work, sorghum sugar
would now be found upon every table in the land, and the
fact that this is not so, but on the contrary that not more
than one operator in a hundred has produced it, under any
circumstances, is directly chargeable to the essentially imperfect mode of treatment which sorghum producers have
been induced to adopt; for nothing is more certain than
that the saccharine matter in the best varieties of sorghum
is almost wholly cane sugar, which may readily be crystallized when the juice is properly defecated, and that *the*

(163)

want of success in crystallizing it, is due the presence of
just those impurities which heat alone cannot remove.

Popular opinion, guided by an experience of more than
six years with the common method of defecation by heat,
and by the representations of persons pecuniarily interested
in securing its continued adoption, is now more unsettled
than ever it has been in regard to the practicability of
sugar production from Northern cane. No progress has
been made in this pursuit commensurate with its importance, or with the results reasonably to have been anticipated from the zealous efforts of the many practical men
that it numbers among its supporters and friends. If we
seek the cause of this condition of things we shall find it
partly, at least, in the injudicious teachings of some who
should know better, leading to misdirected efforts and un-
promising results.

In the face of repeated analyses by the most trustworthy
and eminent chemists, and in spite of convictions which
a candid examination could not fail to impress upon their
own minds, we yet find some advocates of the "heat"
theory strenuously asserting that it is only the inferior, and
practically uncrystallizable grape sugar which sorghum
juice contains, and hence the manifest impropriety of any
further attempts toward advancement on this line.

Others profess to have reached the "ultima thule" of
discovery by sailing in a different direction; they denounce
loudly all "chemicals," even to the use of lime, in saturating the free acid; attribute to heat almost miraculous
powers, and then gravely proceed to inform us that the
talisman whereby this giant of the elements may be aroused
to such unwonted activity is intrusted entirely to their
own keeping, or, in other words, that there is only one
form of evaporating pan upon which heat as an agent in
clarification can be used with success.

The advantage of the auxiliary action of heat in defecation has never been questioned; it is, in fact, wholly indispensable; but it is a great mistake to ascribe to its agency more than is due, and to make that the sole agent which cannot in any case rank higher than a subordinate. Much less warrant is there for the assertion that the purifying virtue of "a simple active heat" is of a superior quality when the evaporating surface is made to take the form of a crooked channel by means of a series of metallic strips placed transversely, extending nearly across the pan from the opposite sides.

No *peculiar* advantage in defecation by this mode is discernible, but the reverse; for each time that the current of juice passes to the side of the pan it is compelled to pass through the dense body of scum settled there, and much of the feculent and mucilaginous matter which had already been separated from the juice at the heated center of any one of the transverse channels, soon becomes firmly incorporated with the same juice again, when the latter, rendered more dense by evaporation, passes onward through it, and this deleterious effect is produced as often as the current is led from side to side. Nor does the novelty of this mode of evaporation more deserve to pass unquestioned than its merits. It may be original, but it is certainly not new. Prof. McCulloh, in his work (*Reports of Scientific Investigations in Relation to Sugar and Hydrometers*, revised edition, p. 237), describing various arrangements of evaporating surfaces, and particularly inclined planes, with a grooved or undulated surface to diffuse and retard the descent of the syrup, says:

" M. Derosne has invented an apparatus of this kind, in which he causes the liquid to traverse the whole surface of the inclined plane, by means of numerous strips of metal placed across, and which are not quite as long as the in-

clined plane is broad. The space between these strips
forms a continuous channel through which the juice flows
alternately in opposite directions, and thus passes slowly
over the whole surface before it arrives at the bottom."

It would be difficult to describe more accurately and
minutely in its essential features the sorghum evaporator
before mentioned, but from the connection in which the
description of this one is found, it is evident that it was
proposed to limit its use to the evaporation of previously
defecated solutions, and therefore that it was not open to
the objection above alluded to.   Derosne's pan was heated
by steam, but this form of surface was not found to possess
any peculiar advantages, nor can any such be justly claimed
for it now in its present resuscitated form.

It is noteworthy, also, that those who advocate the use
of heat, unassisted even by lime, thereby limit its action
to a very small share of the influence of which otherwise
it is really capable.   Fresh cane juice is always more or
less acid, and when in this condition its temperature is
raised to the boiling point, and it is afterward allowed to
remain at rest for a short time, a scum separates, in which
may be detected more or less albumen, green coloring matter, etc.   The liquid underneath is still more or less turbid,
and of a greenish-yellow tinge.   The action of heat is confined to the coagulation of only a portion of the albuminous matter, the remainder is held in solution by means of
the acid, a boiling heat has no further effect upon it, and it
passes into the syrup.   The greenish tinge so common in
all the light-colored acid syrups is the best evidence of the
presence in it also of incoagulable albuminous matter and
gum, which hold a portion of the coloring matter suspended
in the solution.   The consequences of this practice are
hurtful to both the sugar and syrup by 1. The production
of an acid syrup.   2. The destruction of a part of the

cane sugar. 3. Presenting a great barrier to the crystallization of the remainder. 4. Making the evaporation more troublesome on account of excessive foam. 5. Causing subsequent fermentation in the syrup.

Acid sorghum syrup not only always retains more strongly than any other the peculiarly harsh, unpleasant flavor of the green parts of the plant, but also is subject to deterioration, because either by the action of the heat assisted by the impurities during evaporation, or by subsequent transformation induced by the retained albumen, a large part of the sugar really contained in the juice is converted into uncrystallizable or grape sugar, and the presence of this lower grade of sugar in any considerable quantity is an insuperable barrier to the crystallization of the cane sugar which has not suffered transformation. Such syrups also, however dense they may be boiled, become invariably thinner after the winter is past, the result of partial decomposition. Fermentation soon sets in. During evaporation such syrups are more frothy, and the viscidity of the scum renders its separation more difficult.

The solution of albuminous matter in the boiling syrup is productive of further evil effects. It impedes the action of boiling, itself, by preventing the mobility of the particles and the free circulation of the currents in the heated liquid. Such syrup is heated less readily and cools more slowly; more fuel is used to reduce it; the boiling point is higher, and consequently there is much more danger, than in well-clarified syrup, of destruction of the sugar at the last by converting it into caramel. A slimy feculent matter commonly separates from undefecated syrup, collecting upon the bottom of the evaporating pan, which soon hardens and burns upon the metal. This, if not constantly removed, soon destroys the pan, as well as burns and discolors the syrup, which scorches upon the burning crust and beneath it when it cracks open. This crust is of a very different kind from that which is formed when too much lime is used, and is more abundant and difficult to get rid of.

In the production of sugar, the common mode of treatment is never in the least successful, except when the canes have been selected, the juice naturally very pure, and the quantity of acid and albuminous matter which it contains is extremely minute.

CHAPTER XXV.

SUGAR MAKING AT HOME.

Adaptation of the System herein recommended to Operations of
Different Degrees of Magnitude—Outlines of a Method designed
to meet the wants of the Farmer who cultivates and works up
a Crop of Fifteen or Twenty Acres of Cane.

No process of sugar manufacture can be considered
complete, or is well adapted to general introduction, that
cannot be applied to operations of very different degrees
of magnitude. It is neither possible nor desirable here to
detail the various modifications to which varying conditions and circumstances may give rise, but ordinary intelligence and good judgment will readily supply them. A
general plan only can be indicated, the essential features of
the method to be used pointed out, and the apparatus and
means necessary to the attainment of the desired ends described. The growth of this new branch of industry will
doubtless in many places give rise to large establishments,
supplanting to some extent more limited individual operations, yet it must be remembered that it is with the latter
that the success of the whole enterprise is at present iden-
tified.

It was only after many years of trial of mills propelled
by animal power and inexpensive apparatus that the sugar
industry of Louisiana was established upon a solid basis;
and notwithstanding that the most elaborate and expensive
machinery has of late years been in use there, small plant-

15   (169)

ers still adhere to the more simple appliances, and with a very marked degree of success and profit. Here also large works will not interfere with those conducted on a very moderate scale, if the latter be managed with skill and prudence. In fact, the expense of transporting large quantities of cane from different parts of an extended area of country to one great central manufactory is a source of great loss in many respects, which is scarcely balanced by the advantages accruing to a great concentration of capital and skill.

There is a large class of persons in our country having lands well adapted to sugar growing, and possessed of sufficient energy and intelligence, whose means or opportunities do not permit them to engage largely in this pursuit, but who would be glad to have it within their power to work up the cane which they could grow upon their own lands; and it is just this class of persons—farmers of intelligence and energy, desirous of making use of all available means of enhancing their own comfort, profit, and independence, who have thus far, under so many discouraging circumstances, lent their aid to this enterprise, while capitalists have been lagging in the rear.

The question is often asked by such, How can the planter work up to advantage a crop of from ten to twenty acres of cane on his own land and under his own care—conducting the whole series of operations, beginning with the working of the soil, and the planting of the seed, and ending with the production of a good article of brown sugar? The answer that I shall give to this inquiry is based entirely upon my own observation and experience.

By pursuing the method of cultivation referred to in a former part of this work, fifteen to twenty acres of cane can be well attended to by a single hand, from the time of planting until the close of the period of cultivation; or in lat. 40°, from the 20th of April until the close of June, except at the time of the second plowing, when two more persons will be needed to hoe the cane. During this period it will have received one good hoeing, and will have been twice or three times plowed. The ground at that time will be free from weeds, and the cane shooting. It may then be safely left to take care of itself. It will require no more work until September, when the blading, cutting, hauling, and the subsequent process of manufacture will follow in regular order, and require generally the labor of three persons until the cane is all worked up, and then one intelligent man can prepare the sugar and syrup for the market as the different lots come into condition to be successively operated upon.

The evaporating range should be not less than twenty-four feet in length, and the mill should be of the largest size, adapted to horse-power. The rolls should not be less than a foot in diameter, and fifteen inches long, if all of the same size; with the requisite strength in all the other parts to sustain the strain that would ordinarily be applied by the use of four horses. The mill should be placed upon a strong platform of plank, supported by stout timbers. This platform should be built on a declivity, or preferably, when circumstances will permit of it, it may extend from the projection commonly known as the "overshot" of a side-hill barn. The horses work on the ground floor below; the swape is a straight beam secured in a horizontal position at a height suitable for easy draught, to a vertical wooden shaft of ten or twelve inches in diameter, which is strongly coupled to the shaft of the driving roll. Mills with horizontal rolls possess superior advantages when used in this manner, and require an increase of motive power scarcely appreciable in practice.

The convenience of this arrangement is obvious. The horses work to good advantage; the vicinity of the mill is clear of all encumbrance; the loss by waste, dirt, damage

to machinery, etc. is much diminished.   There is room at
one side, and sufficient elevation to allow the trash to be
thrown into the barn-yard either by hand or by means of a
carrier or apron, dumping it over the platform outside the
horse-walk, where it will be in a situation to be disposed
of to the best advantage in its subsequent conversion into
manure. (See Chap. VII.)   On the other side, the barn
floor is on a level with the ground, where the cane is received to be passed through the mill.   Part of the cane
may be piled up outside in a protected situation ; but as
much of it as possible should be stored away in the side
bays of the barn, where it will keep in perfectly good condition, and may be worked up during all the wet and in-
clement weather, when that outside should not be handled.
A supply of cane, sufficient for a day's work, may always
be stacked alongside the mill, on the barn floor, at intervals of leisure, so that the constant attendance of one man
to carry cane to the person who feeds the mill may be dis-
pensed with.   The mill, the horses, the cane, and all the
machinery being thus under roof, there need be no interruption of the work.   The evaporator, etc. should be
placed at one side under cover, about on a level with the
horses, securing thereby an abundance of fall for the juice
through a long pipe into the tanks and pans to the cooler,
and thence into the vessels to contain the sugar or syrup.
The evaporating room should stand at such a distance,
and should be so placed as to avoid all danger to the surrounding buildings from fire.   There is but little danger,
however, from this source, if the furnace is properly constructed, and the smoke-stack is of sufficient height to secure a good draught, and to prevent the escape of sparks.
The evaporating and crystallizing rooms should be contiguous to each other, and a tram-way constructed from
one into the other upon which to convey on a truck the
syrup to be crystallized.

CHAPTER XXVI.

VALUE OF SORGHUM IN SUGAR PRODUCTION.

Capacity of Sorghum for Sugar Production as ascertained by Analysis and Experiment—Analyses by Dr. Jackson, of Boston—Prof. Lawrence Smith, of Louisville—Analyses in France by Madinier and Vilmorin—Extended series of Experiments made by Mr. Jos. S. Lovering, of Philadelphia—Observations upon a Report of Analyses made by Dr. Wetherill in the Laboratory of the Department of Agriculture, at Washington.

IN order to arrive at a true estimate of the value of this plant in sugar production, some established variety of it must be assumed as a standard of comparison : and it is to be regretted that for want of due care on the part of planters in preserving the purity of the cane, no one variety can at present be selected which, making due allowance for differances of soil, climate, and culture, is of an entirely uniform character. This is especially true of the Chinese sorghum. Yet I am convinced, from my own experiments and those of others, that care in preventing hybridization of the cane, and providing for it the conditions of successful cultivation already indicated, improvement in saccharine richness, as well as in the general vigor of the plant, will be the uniform result. Instead of a juice marking in the unclarified state 7° or 8° Beaumé, this variety with me now marks regularly not less than 10°—often 12° —with an average of 11°, or $9\frac{1}{2}°$ when properly clarified, and containing seventeen or eighteen per cent. of saccharine matter—nearly all of which is true cane sugar.

The following I have found to be average results from ripe cane, grown on good upland soil:

Specific gravity of fresh juice at 60° F., 1085 or 11° Beaumé.

Specific gravity of clarified juice at 60° F., 1070 or 9½° Beaumé.

Whole amount of saccharine matter, 17 per cent., which is cane sugar almost exclusively.

How this accords with an analysis made by Dr. Charles T. Jackson, an eminent chemist of Boston, Mass., in the year 1857, will be seen in the following extract from his report. (Ag. Rep. of Patent Office, 1857, pages 187 and 189.) The subject of this analysis was ripe Chinese cane.

Specific gravity of the filtered juice, 1·062. Calculated saccharine matter per cent., 15½. Obtained result, saccharine matter (cane sugar nearly all crystallized), 16·6 per cent.* Some imphees analyzed by him gave from 14·3 to 15·9 per cent. of sugar (well crystallized cane sugar) (see table), which also corresponds closely with the per cent. of saccharine matter obtained from imphee by Mr. Leonard Wray.

In the same Report, p. 194, Prof. J. Lawrence Smith, of Louisville, Ky., publishes an account of an " investigation of the sugar-bearing capacity of Chinese cane," in which he gives the following as the composition of the stalk of the cane:

Water.. 75·6 per cent.
Sugars....................................... 12·0† "
Woody fiber, salts, etc...................... 12·4 "
                                            ———
                                            100·0

* An erroneous statement appears in the Report of the Department of Agriculture for 1862, p. 533 (table), in which the result as obtained by this chemist is stated to be 10·15 per cent. total of sugar, instead of 16·6 per cent. as above.

† This is equal to nearly 14 per cent. *of the juice*, instead of 12 per cent., as it is made to appear on p. 533 of the Report for 1862.

In this twelve per cent. of sugar, an examination by Soleil's polariscope indicated the proportion of the crystallizable to the non-crystallizable sugar to be as 10 to 2. And hence he asserts the opinion that "this result settles the question that *the great bulk of the sugar contained in the sorgho is crystallizable or cane sugar proper.*"

From the above it is evident that the cane juice examined by Prof. Smith contained about 14 per cent. of sugars (13·7), the proportions of cane and uncrystallizable sugars being respectively 11·4 and 2·3 per cent. The juice did not equal in saccharine richness that analyzed by Dr. Jackson, but this difference of quality is not greater than is often met with in canes grown on different varieties of soil, or under other dissimilar conditions; but the statement that the uncrystallizable sugar exists in it in very small quantity relatively, is in accordance with the foregoing results.

These analyses were made at about the same period, the one upon canes grown on the Atlantic slope, the other upon cane of the same variety grown in the Mississippi Valley; and they have a permanent value as indicating the saccharine qualities of this cane at the time of its introduction into this country.

M. Madinier, of Paris, in a letter, an extract of which is published in the *Agricultural Report* for 1856, p. 313, states that the juice of Chinese cane grown in the north of France, in lat. 49°, yielded 16 per cent. of sugars as determined by Clerget's optical instrument. This must be regarded as a very favorable result, inasmuch as the climate in which the cane was grown is scarcely better suited to this plant than that of England—being sufficiently moist, but lacking the fervid glow of sunlight which prevails during the summer months in this country in the same latitude; and it is remarkable that there the percentage

of uncrystallizable sugar was found to be so low relatively
in the juice of canes necessarily but imperfectly matured.

M. Madinier writes : " It is certain that from this plant
crystallizable sugar can be extracted similar in every respect to that made from the cane of the tropics.   Of this
I entertain the highest conviction, which is supported by
authentic, though not very numerous facts. * * * The
stalks of the sorgho contain crystallizable sugar, without
furnishing a greater quantity of molasses than the cane.
An experiment made at Veriéries, with Clerget's apparatus,
showed the juice to contain 16 per cent. of sugar, of which
there were only $10\frac{1}{3}$ per cent. crystallizable, and $5\frac{2}{3}$ per
cent. uncrystallizable.   Yet, we can by no means depend
upon a result gained from plants grown in the Department
of the Seine and the Oise, in a climate altogether beyond
the range adapted to the sorgho."

M. Louis Vilmorin, of Paris, alluding to his own experiments upon canes grown, no doubt, in a more favorable
climate than those of which M. Madinier gives an analysis,
says :* " The crystallization of the sugar of the sorgho, it
seems, should be easily obtained in all cases where the cane
can be sufficiently ripened ; and as the proportion of the sugar
is an unfailing index of ripeness, it follows that we should
always be sure of obtaining a good crystallization of juices,
the density of which exceeds 1·075, while weaker ones could
not yield satisfactory results after concentration.

" I attribute this peculiarity to the fact that the sugar is
preceded in the juice by a gummy principle, which seems
to be transformed, at a later date, for its proportion diminishes in exact correspondence with the increase of the
saccharine matter.

---

* Ag. Rep. for 1856, p. 312.   M. Vilmorin's letter is dated April 20,
1857.

" The uncrystallizable sugar, or glucose, undergoes the same change; that is to say, it is more abundant before than after the complete maturity; but its action seems less unfavorable to the progress of crystallization. The gummy principle obstructs it in two ways; for, besides being a serious obstacle to the commencement of crystallization, it afterward renders it almost a matter of impossibility to purge the crystals, if obtained.

" However, as I observed, this difficulty only presents itself in the employment of unripe canes; *for as soon as the juices attain the density of* 1·080 *and more, they contain but little else than crystallizable sugar*, and their treatment presents no difficulty.

" The lime employed, even to a slight excess, is not so detrimental, it seems to me, in practice, as theory would perhaps indicate. Perhaps a slight fermentation, which is inevitable, may disengage enough carbonic acid to destroy the uncrystallizable compound formed by its union with the sugar. The fact is, that the best crystallizations obtained have occurred in those experiments in which I feared to have used too much lime."

All the authorities above quoted concur in proving the fact that immediately before and after the introduction of the Chinese cane into this country, cane sugar constituted nearly all of the saccharine matter in the ripe plant. A complete practical demonstration, however, of this fact, outside the laboratory, with only such aids as are within common reach, guided by good judgment and ripe experience, was made about the same time, and the results published to the world.

During the autumn of the year 1857, Mr. Joseph S. Lovering, the eminent sugar refiner of Philadelphia, instituted an extended series of experiments upon cane grown upon his own grounds; and from the yield of sugar and

molasses actually obtained by him, he gives the product of
an acre of cane definitely as follows :

	lbs.	galls.
Actual yield, crystallized sugar.............1221·85 Molasses, 74·39		
Add for inefficiency of mill 10 per cent.		
" " reheatings, etc..... 5 "		
" " footings, etc........ 5 "		
Total.......................... 20 per cent.=244·37		
Probable yield per acre. Sugar...........1466·22 Molasses, 74·39		

The actual yield of sugar and syrup as above stated per
acre is, I believe, about a fair average result—such as may
be attained from pure cane planted upon a suitable soil,
from the latitude of Philadelphia southward.   But it will
be observed that the yield of juice was much greater, and
its saccharine richness proportionally much less than ordinarily is to be anticipated. The expressed juice amounted
to 1847 gallons per acre, or about one-third more than is
commonly produced from an equal area ; its density in the
crude state was 10° Beaumé, but only 7½° clarified (at
162 F.), containing about 13 per cent. of saccharine matter, as indicated by its specific gravity, or 12·72 per cent.
the actual yield of sugar and molasses.   Of this, 7·35 per
cent. was crystallized sugar, 5·37 per cent. molasses (drippings). Taking into the account the 10 per cent. of sugar
lost in reheating and footings as above stated, we should
have of cane sugar 8 08 per cent.; molasses, 4·64 per cent.
The molasses, however, still contained an indeterminate
proportion of crystallizable sugar; another proportion of
it consisted of cane sugar necessarily converted into liquid
or uncrystallizable sugar during evaporation: hence the
per cent. of cane sugar originally contained in the juice
must have been considerably greater, and that of uncrystallizable sugar proportionally less than in the above
estimate.

Further, since 16,530 pounds of juice (1847 gallons, weighing 8.95 pounds per gallon) produced but $2114\frac{1}{2}$ pounds of saccharine matter—the product per acre of syrup reduced to the striking point—238° F., could not have exceeded 180 gallons, and hence it required 10 gallons of juice to produce one of such syrup. These results prove that the juice of ripe and unhybridized cane contains a variable amount of sugars and impurities due, in a great measure, to peculiarities of soil and season, and confirm the conclusion, then reached by analysis, that the quantity of uncrystallizable sugar contained in the juice as a natural product, and not formed from the cane sugar by the action of heat, etc., is comparatively insignificant.

Mr. Lovering did not repeat his experiments, his object being simply to test the value of sorghum as a sugar-producing plant. In a recent letter he says: "In the year 1857, with no other object in view than to satisfy my own curiosity, I undertook to ascertain a fact that had been denied by some notable chemist, viz.:—the presence of cane sugar in the juice of sorghum. The use of Scliel's saccharometer very speedily proved the fact that it does contain cane sugar in sufficient quantity to render its extraction profitable, and highly important in our Middle and South-western States. Having ascertained this truth, I was encouraged to proceed a step further, and undertook the experiments, of which I published a detailed account in 1858."*

---

* It is surprising that Mr. Lovering's unasserted but just claim to the thanks of the public for the faithful performance of an undertaking to the success of which he was in no sort pledged, and in which he had only a common interest with all men who desire to know the truth, should in some quarters call forth only disparagement, or subject him to such invidious comparison as is expressed in the following extract from a widely circulated advertising pamphlet:

Since then, occasional instances of successful crystallization have been recorded ; but they stand out as isolated
examples in the midst of many failures ; the results have
been far from uniform — success has been the exception
rather than the rule—and they have been valuable only as
showing the utter futility of the means ordinarily employed.
Encouraged by the fact that sugar has been made from
sorghum, many practical men have spent years in vainly
striving "to learn the knack of it," and failing, have
abandoned it in despair, or catching at straws, have become
the victims of the first "patent recipe" vender that came
their way.

The report of Dr. C. M. Wetherill, late Chemist in the
National Department of Agriculture, is a valuable contribu-
tion to our knowledge upon this subject, but it is unfortu-
nate that the opportunities afforded him for making an
analysis of the best varieties of the new cane were entirely
inadequate.   No analyses were made before the month of
November, when most of the samples of cane, some of
them transported several hundred miles, were, as is inti-
mated, more or less injured.   Some of the specimens were
immature, others were undergoing fermentation, and hence
the examination of them was necessarily hurried, and under

"A RECOLLECTION.

" In 1857 two men were at the same time laboring to develop the
merits of the sorgo : one in a Philadelphia refinery with all the appliances capital could confer; the other a backwoods chemist, with nothing
but an old skillet and a cooking stove. The experiments of the former
were very elaborate and expensive, *but there they ended.* The latter,
ridiculed by his neighbors as a fool for thus wasting his time, sat pa-
tiently over his skillet, studying into the nature of the juice as developed
before him, and clutching every new idea. until after many failures in
his efforts to embody the suggestions made by his study, he succeeded in
bringing out that remarkable and justly celebrated boiler known as
Cook's Evaporator," etc.

these circumstances were regarded as merely preliminary to future investigations of a more extended and valuable character.

The discordant results of the different analyses, however, after making due allowance for all assigned influences, as well as those peculiarities induced by differences of climate and soil, reveal the unwelcome fact of which there was previously some evidence, that the great bulk of the sorghum now grown in this country is of an inferior quality to that first introduced. This misfortune is the result of the hybridization of the cane with broom corn, and deterioration by improper culture, the consequences in most cases of ignorance of the nature and wants of the plant. This discovery is of itself sufficient to impair the value which these analyses would otherwise have possessed as affording a fair index of the saccharine quality of the cane in its native purity. While there is yet in this country undeteriorated cane, the juice of which uniformly contains not less than 16 per cent. of saccharine matter, which is cane sugar almost exclusively, none of the specimens sent to the laboratory at Washington contained more than 10 per cent. of cane sugar, and some of them not more than 2 or 3 per cent.; the whole amount of sugars of both kinds rarely equaled 15 per cent., and in some canes it fell as low as from 5 to 7.

The report upon the canes examined is valuable as an exhibit of the present condition of the plant in this country, but as an index of its true value for the production of sugar, it is not to be relied upon. Nor for the determination of this important point are the analyses of sugars and syrups there presented of much greater consequence, however skillfully and conscientiously they may have been made. Nearly all the samples of syrup exhibited a precipitate of crystals of cane sugar in the vessels which contained them.

In some cases the mass of crystals deposited equaled in volume the clear syrup above them from which they had separated. For the reason that the condition of the syrups at the time they were sent to the laboratory at Washington could not, in many cases, be ascertained; in the analyses no account is made of the sugar which had been deposited in this way; but in the table inserted in the Report only the amount of cane and uncrystallizable sugar in solution at the time the analysis was made, is made to appear. The necessity which led to this arrangement is to be regretted, for some true syrups of drainage (molasses) from which the greater part of the sugar had already separated are here compared side by side with others which were known to contain all their original sugar. But it is worthy of note that specimens which had previously deposited a mass of crystals equal in bulk to half the length of the bottle which contained the syrup, still contained a very large per cent. of cane sugar,—in one instance, 35·62 per cent. of cane and 28·11 per cent. of uncrystallizable sugar.*

---

* No. 7. White imphee syrup, uncryst. sugar, 28·11 per cent.; crystallizable sugar, 35·62; water and impurities, 36·27 = 100·00. If to the 35·62 per cent. of cane sugar, we add the "sediment of light-colored crystals of cane sugar extending half the length of the bottle," which may be inferred to have amounted to near 50 per cent. of the whole contents of the bottle by weight—say 45 per cent., the 55 per cent. of clear syrup, analyzed, contained:

Uncrystallizable sugar	15·46
Cane sugar	19·59
Water and impurities	19·95
	55·00
To which add cane sugar crystallized . . .	45·00
	100·00

Hence this syrup originally contained 64·59 per cent. cane sugar, and 15·46 uncrystallizable. It may be inferred that a part of the 15·46 per cent. is sugar reduced to an uncrystallizable condition by the action of heat. The original per cent. of uncrystallizable sugar must have been exceedingly small and the juice of great purity.

It is probable that in all cases in which the cane from which these syrups and sugars were made was pure, that nearly all the uncrystallizable sugar left in the molasses was decomposed cane sugar, resulting from defective means of evaporation, want of proper defecation, etc. This is now known to be true as respects the uncrystallizable sugar in the syrup made from Southern cane, although it was long thought to be a natural product, and treated as such.

The samples of sugar sent to Dr. Wetherill were unequivocally cane sugar, and in most cases exhibited a degree of purity equal to that of commercial sugars made from the tropical cane and the beet.

The results of all the foregoing analyses and practical experiments, corroborated by my own investigations, made with care, during a series of years, fully justify me in asserting that there is yet unhybridized Chinese cane in this country, containing in the matured juice not less than 16 or 17 per cent. of saccharine matter, which is nearly all crystallizable cane sugar.

CHAPTER XXVII.

HOW TO TEST CANE JUICE.

Importance to the Manufacturer of some Means by which he may learn the Value of any given Sample of Cane Juice—Characteristics of Crystallizable and Uncrystallizable Sugars—Microscopic Sections of the Stem of Chinese Sorghum and Imphee—How to distinguish between Grape and Cane Sugar in the Solid Form—Methods of determining approximately the Relative Proportions of Cane and Grape Sugars, and Impurities in Sorghum Juice—The Hydrometer—Subacetate of Lead—Modification of Fehling's Copper Test—Peligot's Test by Sugar Lime.

It is possible, by means of certain chemical processes, to determine accurately the composition of the cane juice, and to indicate the precise quantities of crystallizable and uncrystallizable sugars in any given sample. To the sugar grower it is always of the first importance to have at hand the means of ascertaining this, if not with rigorous accuracy, at least to within a close approximation to the truth. Often a simple experiment directed to this end, if carefully performed by him, will serve as a guide to the mode of treatment to be pursued in any given case. It is not necessary that he should previously be possessed of any chemical knowledge; he has simply to become acquainted with some of the characteristic properties of the substances which are daily passing through his hands. Without this he cannot have a competent knowledge of his art. A preliminary test, requiring but little time or care, would often

(184)

be invaluable to the manufacturer, if he knew how to apply
it, in order that he might avoid errors, in which much time
and money would be thrown away.   It is true, that by
adopting a proper system of cultivation, the cane may be
made to yield a juice of a pretty uniform character, yet we
never can have the product under perfect control; the
quality of the juice will vary within certain limits, with
the variable influences of soil, season, situation, and other
conditions.

It is desirable, oftentimes, to know what particular changes
may have been the results of certain methods of culture, or
of the use of special fertilizers.   Improvement in the saccharine quality of the juice, and in the health and vigor
of the plant, are objects of constant solicitude to the
planter: but without a criterion by which to mark his
progress, how can he know whether he is doing well or ill?

The characteristic properties of the two kinds of sugar
usually found in sorghum juice, here first demand attention.
In the ripe, unadulterated plant, the uncrystallizable sugar
is found only in minute quantity, its presence being manifested only by the use of very delicate chemical tests.   In
ripe cane, which has hybridized with an inferior variety,
or which has been grown upon unsuitable soil, or exposed
to other improper influences, the disproportion between
the two kinds of sugar is very much diminished, and the
juice is more impure.   In *unripe cane*, grape sugar is
largely predominant, and is often the exclusive product.
It seems to be impossible to separate this sugar, along with
the cane sugar, in the solid form, although a part of it may
be made to assume a granular condition when cane sugar
is not present, after evaporation and a long period of rest.
Another part, however, permanently retains the liquid form
under all circumstances, and this has been denominated
fruit sugar, but a solution containing cane sugar is capa-

16*

ble of yielding in the solid or crystalline form cane sugar
only.   Inasmuch as the other two varieties of sugar above
mentioned are not distinguishable from each other by the
ordinary chemical tests, and exhibit the same properties,
they are recognized throughout this volume by one name,
that of grape sugar.   They form a viscid liquid, which, if
found in large proportion in syrup, renders the cane sugar
practically uncrystallizable, by acting as a barrier to the
union of its particles.

An incontrovertible evidence of the presence of cane
sugar almost exclusively in the juice of sorghum, is afforded
in the fact that thin sections of the fresh stalk of the plant
under the microscope exhibit the cells filled with innumerable minute crystals of pure white sugar, which by their
form and other criteria are shown to be cane sugar only.
Scarcely a trace of any other substance is found in the cells.
This is well represented in the engravings.

Fig. 3 represents a thin slice of the stalk of Chinese
cane (transverse section, from near the center of the stem),
magnified 200 diameters.

Fig. 3.   Fig. 4.

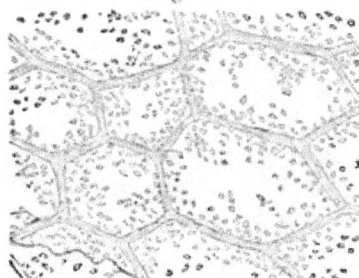 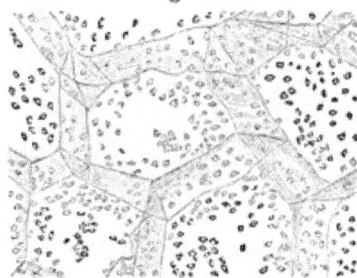

Fig. 4. Transverse section of Nceazana or white Imphee,
magnified 200 diameters (outer series of cells).   The cells
in this variety of cane are very large.

The pithy part of the stalk throughout, uniformly presents the same appearance as that given above.

IN THE SOLID FORM cane and grape sugar differ in many particulars.

1. *Crystalline form.* Cane sugar occurs in the form of bold, angular, colorless crystals, having the form of oblique six-sided prisms, when a hot saturated solution is allowed to cool and evaporate slowly. Crystals rapidly formed are generally imperfectly shaped, but present various modifications of the regular form.

Grape sugar, on the contrary, takes the form of minute tubercular or wart-like granules (*sucre mamelonne* of the French), which, under the microscope, assume the appearance of delicate feathery tufts. There is no resemblance between the regularly crystallized, brilliant, prismatic cane sugar and grape sugar, as will be evident if we compare good refined sugar with the white granules on the surface of raisins, or the mealy sediment in honey, or in jellies prepared from acid fruits, such as the American crab-apple or currant.

2. *Specific gravity.* Cane sugar has a much greater specific gravity than grape sugar. The following are average results.

Specific gravity.
Cane sugar (Rock candy, Pereira).........................1·606
Grape sugar, crystalline tufts, variable
   with the dryness...................................1·390 to 1·400

3. *Solubility.* One ounce of water at ordinary temperatures, can dissolve three ounces of cane sugar, but only two-thirds of an ounce of grape sugar.

4. *Sweetness.* The sweetening power (and consequently the value) of cane sugar is much greater than that of grape sugar, one pound of the former being equal to two and a half pounds of the latter.

5. *Fusibility.* Pure crystals of cane sugar, heated to a temperature of 356° F., melt (impure sugars at a lower

heat, Muscavado at 280° F.); at a higher degree of heat (about 400° F.) they decompose, losing two equivalents of water, and are converted into *caramel*, a dark-brown substance possessing an alkaline reaction, of the chemical composition $C_{12} H_9 O_9$, and freely soluble in water and alcohol. It is from this substance that most of the dark coloring matter in molasses is derived.

Grape sugar when a dry solid (specific gravity 1·39) fuses into a thin watery liquid at a temperature of only 220° F. When heated to 270° F., this fluid sugar boils briskly, giving off $\frac{1}{10}$ of its weight of water, and concretes on cooling, into a bright-yellow, brittle, deliquescent mass.

6. *Deliquescence.* Pure cane sugar remains unaltered in the air ; grape sugar absorbs moisture, and becomes wet and clammy (impure cane sugar also).

7. *Reaction with sulphuric acid.* Cane sugar, when acted upon by strong sulphuric acid, assisted by a slight heat, becomes rapidly blackened (caramelized).

Grape sugar dissolves freely in the acid without blackening.

8. *Reaction with alkalies.* When cane sugar is heated with a little of a strong solution of potash or soda to the boiling point, the liquid does not become colored, but grape sugar similarly treated assumes a brown tint.

9. *Chemical composition.* These sugars are composed of the same elements, but they are not united in the same proportions.

Cane sugar is...$C_{12} H_{11} O_{11}$
Grape sugar, dried at 212° F., is ....$C_{12} H_{14} O_{14}$

10. *Relative proportions of carbon.* From the above it is evident that grape sugar contains relatively less carbon. The following were found to be the relative propor-

tions of carbon in different samples of sugar analyzed by
Prout :

Per cent. of Carbon.

Pure sugar candy, or best refined sugar................42·85

East India sugar candy.....................................41·90

East India raw sugar, dry and of low quality.........40·88

Sugar from Narbonne honey...............................36·36

Sugar from starch..36·20

In a solution of sugar most of the tests above given
cannot readily be applied. Only such means can be employed as will not disturb the delicate equilibrium of the
chemical forces upon which the existence of the sugar
depends. The per cent. of sugar in a pure solution may
be most readily ascertained by taking its specific gravity,
and this is the use of the hydrometer or saccharometer.
But this instrument can give correct indications in a pure
solution only. If other substances than sugar are in the
solution, they will of course affect its density, and a false
indication will be the result.

Moreover, in a pure saccharine solution, composed of cane
sugar mixed with grape sugar, the hydrometer does not
enable us to determine their relative proportions. According to Ure, " At 1·342 syrup of cane sugar contains 70 per
cent. of sugar; at the same density syrup of glucose or
grape sugar contains 75½ per cent. of concrete matter, dried
at 260° F., and therefore freed from the 10 per cent. of
water which it contains in the granular state." These sugars
thus differ from each other in the relative densities of solutions containing equal weights of each respectively. Hence,
in even a completely defecated sample of cane juice, the
indications of the hydrometer are of little value, for they
give us no means of distinguishing between the different
kinds of sugar present in the mixed solution. When the
per cent. of grape sugar in a liquid is already known by

other methods, the hydrometer can then be employed to advantage for determining the per cent. of cane sugar, and it should generally be limited to this use.

The hydrometer of Beaumé is the common form, but the scales of different instruments do not always correspond, and hence the continued liability to error. Every instrument before being used should be subjected to the following test: Place the hydrometer in pure water at a temperature of 60° F., and the point near the top of the stem to which it sinks is the 0 of the scale. Prepare a solution of 15 parts of common salt in 85 parts of water by weight, and at the same temperature as the above, and the place to which it sinks should mark 15° of the scale. The value of each degree of this scale expressed in specific gravities, and correspondent per cent. of sugar are given in table II.

A specific gravity bottle affords a convenient means of testing the accuracy of the hydrometer, and when an extreme degree of accuracy is required, it should be used in its place. It is a bottle containing just 1000 grains, or 100 grammes of distilled water at 60° F., when the stopper is inserted and the outside of the bottle is wiped dry. The stopper is of ground glass with a hole through its center, or a groove cut in its side with a file, to admit of the escape of any superfluous liquid in filling the bottle. When such a bottle cannot be had, a phial of any capacity with a similar stopper may be used, and the exact weight of water at 60° F. which it will hold, ascertained   The specific gravity of the liquid to be examined is obtained by filling the bottle with it and dividing its weight by the weight of the water.

Thus if the weight of the bottle full of water at 60° F. is 1000 grains, and the weight of the solution of sugar at 60° F. is 1083 grains, the specific gravity of the solution will be 1083 divided by 1000, or 1·083. Referring to table

II., we find that the specific gravity 1·083 corresponds to 11° Beaumé's hydrometer, or to 20 per cent. of cane sugar.*

The saccharometer is a hydrometer upon which the per cent. of sugar may be read off directly, without reference to a table, and this form should be in general use.

The following tests may be applied when it is desired to determine *approximately* the relative proportions of cane and uncrystallizable sugars, and impurities in a given sample of cane juice. To a given weight of the fresh juice, filtered through a cloth to remove fragments of pith, etc., and poured into a narrow jar or glass test cup, add carefully a few drops of a strong solution of subacetate of lead, stir it with a glass rod, and let it remain at rest until the upper portion of the liquid becomes clear. Decant a little of this into a small phial or test tube, and add a single drop of the solution of subacetate of lead. If the liquid becomes clouded as before, return to the test jar what was taken out, and add a little more of the lead solution. Let the precipitate settle, and test again, continuing to add the subacetate in small portions at a time, until it ceases to disturb the transparency of the sugar solution.† Take a wide tube of about

---

* Since the refractive power of a body in solution is not influenced by chemical combination, it is possible that a more valuable instrument than the hydrometer might be constructed, which would indicate the per cent. of sugar in solution.

† Subacetate of lead possesses the valuable property of separating from the saccharine solution in the insoluble form, any other organic substances which it contains. In order to determine the quantity of impurities, therefore, throw the whole of the precipitate obtained as above, upon a filtering paper, wash it with pure water upon the filter, allowing the water to fall upon it in a fine stream. Repeat this again after the water has passed off, let the precipitate remain upon the filtering paper until it is nearly dry, remove it carefully from the paper with a thin-bladed knife, place it in a thin capsule of silver or copper the exact weight of which is known, and then dry it perfectly in a water or steam bath. Weigh the capsule and its contents, and subject it to a heat strong

a foot in height (a large lamp chimney will answer the purpose), tie a piece of coarse canvas over one of its open ends, and pour in dry. fresh, well washed animal charcoal (such as is used in the filter of the evaporator), until it is about half full, pack it down closely, and then pour in a convenient portion of the clarified juice from the test jar, collect it as it passes through the filter, rejecting the first that passes through, or about one-fourth of the whole volume of the liquid, because it is below the average strength. When the filtration is complete, return the remainder of the liquor to the filter, and filter a second time. If cane sugar only is present, the specific gravity bottle or hydrometer will give its per cent. in the filtered juice. Any excess of lead is retained by the boneblack, and separated entirely from the liquid.

It remains to ascertain the relative proportions of the grape and cane sugars in the solution. Of the different methods that may be employed for this purpose, but one can be recommended to those unprovided with chemical apparatus or unskilled in their use. This method consists simply in evaporating, in a shallow vessel, a weighed portion of the filtered juice, above mentioned, to the consistence of a thick syrup—the evaporation, at least during its last stages, being accomplished by means of the water-bath, or by exposing the evaporating dish to the heat of the steam escaping from a vessel of boiling water. The syrup should then be placed in a situation where it will be exposed to a uniform temperature of about 70°–80° F. If it has not crystallized after the lapse of a week, add to it a few drops

---

enough to burn off all the organic matter. Weigh the protoxide of lead which remains, and its weight subtracted from the weight of the dried contents of the capsule already known, will give that of the organic matter with which it was combined. This method is sufficiently accurate for ordinary purposes.

of alcohol, and let it remain a week longer. If the juice is of good quality, crystallization will rapidly ensue at the first, the whole mass of syrup becoming solidified apparently into crystals of cane sugar. If the proportion of uncrystallizable sugar is large, crystallization will be retarded, and if the experiment was made upon the juice of impure or very unripe cane, it will refuse to crystallize at all.

When the whole mass has become solid with crystals, place it in a bottle, after breaking it up into fragments, and pour upon it four or five times its volume of anhydrous alcohol, digest in a water-bath for half an hour or more, meanwhile shaking the bottle repeatedly. Afterwards let it remain at rest for several hours, and then decant the clear solution. Repeat this washing with alcohol once or twice, then dry the undissolved sugar contained in the bottle, over boiling water, and weigh it. The loss in weight is the amount of uncrystallizable sugar dissolved by the alcohol.

The practical sugar maker may find it inconvenient, or unnecessary to his purpose to continue the experiment farther than to discover to what extent the sugar is crystallizable.

Other and more intricate methods, leading to more immediate and accurate results, will commend themselves to those whose opportunities and acquirements befit them for the task of conducting them with success.

1. THE COPPER TEST. *Fehling's Solution.* — This method depends upon the property possessed by grape or fruit sugar—but not by cane sugar—of reducing to the state of suboxide the hydrated protoxide of copper when the latter is presented to it in an alkaline solution, and the temperature of the mixture is elevated to the boiling point. The quantity of the oxide of copper reduced is proportional to the quantity of grape sugar in the solution,

but uniform results are not obtained except in a nicely
regulated alkaline solution. *Fehling's solution* is of this
character, and is not liable to decomposition at ordinary
temperatures. It may be prepared as follows :

40 grammes of sulphate of copper, 160 grammes of
neutral tartrate of potash (or 200 grammes of tartrate of
soda) are dissolved and added to 700–800 c. c. (cubic centimeters—grammes) of caustic soda specific gravity 1·12.
This dilute with water to 1154·5 cubic centimeters. Of
this solution

$$1 \text{ cubic centimeter} = \begin{cases} 0 \cdot 0050 \text{ grape sugar} \\ 0 \cdot 0045 \text{ cane sugar} \end{cases}$$

or grains instead of grammes—and then 1 grain=0·0050
grape sugar, without further change of calculation. And

$$\left. \begin{array}{l} 100 \text{ parts of grape sugar} \\ 95 \text{ parts of cane} \quad `` \end{array} \right\} = \begin{array}{l} 220 \cdot 5 \text{ CuO or} \\ 198 \text{ Cu}^2 \text{ O*} \end{array}$$

Fresenius, in his late able treatise upon Quantitative
Chemical Analysis, gives very minute directions for the
successful application of this test, which, somewhat condensed, I insert below. It will be observed that sorghum
juice, being a mixed solution of sugars, must be subjected
to two experiments. One portion of a given sample to
determine the per cent. of grape sugar—and another to
determine that of cane sugar by reducing the latter
to the condition of grape sugar, and applying the same
test. By subtracting the quantity of grape sugar indicated
by the first experiment from that indicated by the second,
we obtain the quantity of grape sugar into which the cane
sugar was converted, and thence by a simple calculation
the cane sugar itself, from the data given above.

The cane juice to be tested should be a clear solution,
prepared by precipitating with subacetate of lead, and

---

* Dr. Ure.

filtering through boneblack as before recommended, or by treating about 15 c. c.* of the crude boiling juice with a few drops of milk of lime, filtering through animal charcoal, washing the precipitate thoroughly on the filter, adding the washings to the filtrate, and diluting it to 15 or 20 times its original volume. Add 12 drops of dilute sulphuric acid (SO_3 $HO+5$ water) and boil the mixture from 1 to 2 hours, adding water as it evaporates. This operation is best conducted in a steam-bath. Neutralize the free acid by means of a dilute solution of carbonate of soda.

The sugar solution must be *highly dilute*, containing only one-half, or, at most, 1 per cent. of sugar. If in a first experiment the sugar solution is too concentrated, dilute it with a definite quantity of water and repeat the experiment.

The copper solution prepared, as directed by Fresenius, gives very accurate results.† "Dissolve exactly 34·632 grammes of pure crystallized sulphate of copper, completely freed from adhering moisture by pulverizing and pressing between sheets of blotting paper, in about 200 c. c. of water. Dissolve in another vessel 173 grammes of perfectly pure crystallized tartrate of soda and potassa in 480 c. c. of pure solution of soda of 1 14 sp. grav. Add the first solution gradually to the second, and dilute the deep blue clear liquid exactly to 1000 c. c. Every 10 c. c. of this solution contains 0·34632 grm. of sulphate of copper, and correspond exactly to 0·050 grm. of anhydrous grape sugar. Keep the solution in a cool dark place, in well-stoppered bottles filled to the top, as absorption by carbonic acid would lead to the separation of suboxide of

---

* c. c. cubic centimeters. See table V.

† Fresenius's Quant. Chem. Analysis. London, 1860, pp. 576-9.

copper upon mere exposure to heat. This might be prevented, however, by the fresh addition of solution of
soda. Before using the solution, boil 10 c. c. of it for
some minutes, by way of trial, with 40 c. c. of water, or
dilute solution of soda if there is reason to believe that
the fluid has absorbed carbonic acid; if this operation
produces the least change in the fluid, and causes the separation of even the smallest quantity of suboxide, the solution is unfit for use.

"*The process.*—Pour 10 c. c. of the copper solution into
a porcelain dish, add 40 c. c. of water, or very dilute solution of soda if required, heat to gentle ebullition, and
allow the sugar solution to drop slowly and gradually into
the fluid from a burette or pipette divided into $\frac{1}{10}$ c. c.
After the addition of the first few drops, the liquid shows
a greenish-brown tint, owing to the suboxide and hydrated
suboxide suspended in the blue solution. In proportion as
more of the sugar solution is added, the precipitate becomes more copious, acquires a redder tint, and subsides
more speedily. When the precipitate presents a deep-red
color, remove the lamp, allow the precipitate to subside a
little, and give to the dish an inclined position, which will
enable you readily to detect the least bluish-green tint. To
make quite sure, however, pour a little of the clear supernatant liquid into a test tube, add a drop of the sugar
solution and apply heat. If there remains the least trace
of salt of copper undecomposed, a yellowish-red precipitate
will form, appearing at first like a cloud in the fluid. In
that case pour the contents of the tube into the dish, and
continue adding the solution of sugar until the reaction is
complete. The original amount used of the solution of
sugar contains 0·050 grammes of anhydrous grape sugar.

"When the operation has terminated, ascertain whether
it has fully succeeded; that is, whether the solution really

contains neither copper, sugar, nor a brown product of the decomposition of the latter. To this end filter off a portion of the fluid while still quite hot. The filtrate must be colorless (without the least brownish tinge). Heat a portion of it with a drop of the copper solution, acidify two other portions, and test the one with ferrocyanide of potassium, the other with sulphuretted hydrogen. Neither of these tests must produce the slightest alteration. If the fluid contains a perceptible quantity of either oxide of copper or sugar, this is a proof that too much or too little of the latter has been added, and the experiment must accordingly be repeated. The results are constant and very satisfactory. Bear in mind that the solution of sulphate of copper must always remain strongly alkaline; should the sugar solution be acid, some more solution of soda must be added."

Second method.—This may be resorted to in cases in which from the dark color of the saccharine fluid it is difficult to determine the exact point at which the process of reduction and separation is accomplished. In this case the solution of copper may be used in excess, and the suboxide which precipitates determined.

"This requires the same solutions as the first. Pour 20 c. c. of the solution of copper and 80 c. c. of water, or of highly dilute solution of soda, if required (or a larger quantity of the copper solution diluted with water, or solution of soda in the same proportion), into a porcelain dish. Add a measured quantity of the dilute sugar solution, but not sufficient to reduce the whole of the oxide of copper, and heat for about 10 minutes on the water-bath. When the reaction is completed, wash the precipitated suboxide of copper by decantation with boiling water. Pass the decanted fluid through a weighed filter, dried at 212° F., then transfer the precipitate also to the filter, dry at 212° F., and

17*

weigh.   Or ignite the suboxide of copper with access of
air, and convert it completely into oxide by treating with
fuming nitric acid.

"100 parts of anhydrous grape sugar correspond to
220·05* of oxide of copper, or 198·2 of suboxide of copper,† or 155·55 of iron converted from the state of sesqui-
chloride to that of protochloride.   In the application of
this method, it must be borne in mind that the separated
suboxide of copper will, upon cooling of the supernatant
fluid, gradually redissolve to oxide, being reconverted into
this by the oxygen of the atmosphere.   Hence the necessity of washing the precipitate by decantation with boiling
water."

The details of this method are thus given at length, for
the reason that it is the test most depended upon for determining the quantity of grape sugar in a solution.   With
the exception, perhaps, of that given below, it is the only
purely chemical process known that may be implicitly relied
upon for its accuracy, and by means of which the result is
reached with facility and dispatch.

An elegant quantitative test for cane and grape sugar has
been proposed by M. Peligot, which I have found to give
uniform results.   I give this method as described by Dr.
Ure.   "Peligot's method depends upon the definite constitution of sugar lime (or saccharate of lime), its greater
solubility in water than in lime alone, and the unalterability
of this solution by heat.   Soubeiran found sugar lime to
consist of three equivalents of lime to 2 equivalents of

---

* Fehling obtained as highest result 219·4 grammes of oxide of copper.

† Neubauer found in his experiments with starch that 0·05 of the latter
correspond to 0·112 of suboxide of copper.   As 90 of starch gives 100 of
grape sugar, 0·05 of the former correspond to 0·0555 of the latter.   Accordingly 100 of grape sugar gives actually 201·62 of suboxide of copper,
instead of 198·2.

sugar; *i.e.* 84 parts lime to 342 sugar, or about 1 to 4. 10 grammes of sugar dissolved in 75 cubic centimeters of water, ground up with 10 grammes of slaked lime, filtered and again filtered through the lime, 10 c. c. of the filtrate diluted with 2 to 3 deciliters of water and tinctured with a little litmus, are carefully neutralized with a measured volume of dilute sulphuric acid (21 grm. oil of vitriol in 1 liter water), and the quantity of acid used noted. It gives the quantity of lime neutralized, and, from the above proportion, the quantity of sugar present.

"If cane sugar is to be examined for starch or grape sugar, one test is made as above, and another in which the liquid is heated to 212° F., and then, when cool, tested with the acid. The lime solution with cane sugar becomes cloudy by heat but clarifies on cooling, while if grape sugar be present it becomes brownish yellow, and requires much less acid for neutralization. Indeed, a deciliter of starch sugar solution requires 4 c. c. of the test acid, or just as much as lime-water itself.

"The amount of sugar in a solution is estimated by the amount of lime which it will dissolve, and the lime is determined alkalimetrically by means of the acid. A table has been constructed by Peligot for calculating the results." (See table IV.)

CHAPTER XXVIII.

COMPARISON OF SOUTHERN CANE AND SORGHUM.

1. Botanical Characters—Period of Growth—Propagation—Tillering—Climate—Soil—Manures—Maturity of Juice—Condition of the Juice in different parts of the Stalk—Chemical Composition of the Stalk—Chemical Composition of the Juice.

THE following brief but comprehensive sketch of the chief characteristics, respectively of Southern sugar cane and sorghum, will, it is hoped, serve to correct misapprehensions existing in the minds of many in regard to the relative importance of the latter as a sugar plant. A comparison is here instituted which will show that these two species of plants, although differing widely from each other in some particulars, are really very much alike—that the Northern cane is a true analogue of the cane of the South, in its most important characters, although of a different species—and that in saccharine richness the one approaches much more closely to the other than has generally been believed.

It is obvious that where the resemblance between them is close—due allowance being made for differences in their natural relations to soil, climate, etc.—those modes of treatment, which the experience of two or three centuries has proved to be the best for the one plant, may safely, and with great advantage, be applied to the other. On the

(200)

other hand, where strong, distinctive habits and properties are manifested by each, necessity requires the employment of means and processes adapted to those peculiarities.

1. BOTANICAL CHARACTERS.—Both plants are members of the natural family *Gramineæ—the grasses*—to which also maize, millet, rice and our common cereal grains are referred. Unlike the latter, however, but like maize, they have solid pithy stems, charged with a saccharine juice.

Southern Sugar Cane, or Saccharum Officinarum.

GEN. CHAR.—*Spikelets* all fertile, in pairs, one sessile, the other stalked, two-flowered, lower floret neuter, with 1 palea, the upper hermaphrodite, with 2 paleæ. *Glumes* 2, membranous, subequal, concave. *Paleæ* thin, transparent. *Stamens* 1—3, styles 2, stigmas feathered with simple toothletted hairs.

SPE. CHAR.— *Culm* solid with pith, closely jointed (joints about three inches or more apart), about 1½ inch diameter, brittle, of a green hue, verging to yellow at maturity, 8—15 feet high in Louisiana, sometimes 20 feet in the tropics. *Leaves* flat, linear lanceolate, 3 or 4 feet long, 1 to 2 inches broad, of a sea-green tint, striated, fall off as the plant matures. *Panicle* 1 to 2 feet long, pyramidal, of a gray color from the long, white, silky hairs that surround the flower.

There are several distinct varieties, such as the common yellow or *Creole* cane, the *purple*, *giant*, and *Tahiti* or *Otaheitan.* All, except the last, native in Southeastern Asia. The Otaheitan cane was obtained from Otaheite, one of the Society Islands.

Sorghum.—Sorghum Saccharatum.

GEN. CHAR.—*Spikelets* in twos or threes on slender branches, middle spikelet complete, two-flowered, the lower flower abortive, lateral spikelets sterile, pedicels smooth or slightly pubescent, awned, in some varieties, bearded or downy.   *Glumes* 2, coriaceous, concave—paleæ membranous.   *Stamens* 3, styles 2, stigmas feathered.

SP. CHAR.—*Culm* solid with pith, joints 10 inches apart in some varieties, in others not more than 3 to 7 inches (Liberian), 1 inch to 1½ inch in diameter, 8 to 15 feet in height, green—some varieties yellow when mature. *Leaves* lanceolate, acuminate, pubescent at the base, 2 to 4 inches broad, persistent. *Panicle* somewhat diffuse (Red imphee and Chinese), or compact and erect (Oomseeana), or drooping at top when mature (Neeazana).

Numerous varieties.   Chinese cane derived from the north of China.   Imphee cane from South Africa, near Cape Natal, in Caffer-land.

2. PERIOD OF GROWTH.   *S. cane.*—Perennial from the root-stalk in the tropics, flowering in from 12 to 20 months. Period from planting to flowering 12 months, from planting to ripening 16 to 20 months.   In Louisiana it never matures its seeds in ordinary cultivation.

Sorghum, annual, at least in temperate latitudes.   The period of growth differs according to the variety, varying from 3 to 5 months.

3. PROPAGATION.   *S. cane.*—Propagated, ordinarily, in tropical climates, and always in the sugar district of the Gulf States, from cuttings—joints, 15 to 20 inches in length, taken from the top of the stem, which is the least valuable part of the plant.   In Louisiana, on account of constant deterioration, it is necessary to replant every third

or fourth year, twenty-five acres out of every hundred generally being constantly employed in the propagation of the joints from which the cane on the other seventy-five acres is grown.

Sorghum. — Propagated from seeds planted annually. In this case all the labor is saved which, in the cultivation of the Southern cane in Louisiana, is expended in working the unproductive fourth of the whole breadth of land.

4. TILLERING. *S. cane.*—The cutting or buried joint throws out a number of stalks (ratoons) in a manner analogous to that by which a single root of wheat becomes multiplied, *i.e.* by *tillering.*

Sorghum.—In a similar way, as already shown (see Ch. XI.), the crop of sorghum is largely increased by side shoots arising from the root. Sorghum cannot be propagated from cuttings of the stem, but when the stems are cut down they ratoon, giving a second crop, which does not ripen except in climates where the summer is sufficiently long. Mr. Leonard Wray states, that in South Africa he has grown ratoons of Neeazana, or white Imphee, six feet high, and in flower in two months after the first cutting,—sometimes fifteen stalks tillering out from one root. These ripened their seed.

5. CLIMATE. *S. cane.* — The production, within the United States, restricted to a very narrow belt of country bordering on the Mexican Gulf.

Sorghum.—Already defined (see Ch. IX.); some varieties can be ripened wherever Indian-corn is successfully grown.

6. SOIL. *S cane.*—The most favorable soil is a deep, rich, moist loam. The soil of new lands and of rich valleys is pernicious to the saccharine quality of the juice. The sugar is always dark when the cane is grown in such situ-

ations, and the juice is poor, "requiring double the quantity to the hogshead that it does when the canes are of less rank growth."*

Sorghum.—More sensitive in this respect than Southern sugar cane. The same kind of influence is exerted upon the juice by the soil of new land, but in a much higher degree. The Chinese sorghum is more liable to injury from this source than the imphees. The most suitable soil is deep, rich, elevated, and calcareous. (See Ch. VIII.)

7. MANURES. *S. cane.*—When animal manure is largely supplied to the plant, the ammonia augments in the juice the substances containing nitrogen (albuminous and glutinous matter), and hinders or totally prevents crystallization of the sugar. The same effect follows in the South after rotation with the *cow-pea.*†

Sorghum.—Exhibits very markedly in the character of its juice, exposure to similar influences. The ammonia sometimes imparts its taste to the syrup from cane largely supplied with barn-yard manure.

8. MATURITY OF THE JUICE. *S. cane.*—Juice matured, or of the greatest saccharine richness before the time when the plant is in flower, during which time and afterward its yield of cane sugar rapidly diminishes, the latter being transformed into other substances, starch in the seed, or woody fiber in the stem.

SORGHUM.—*Juice matures after the flowering process has passed*, and when the seed is hardening (imphee) or perfectly ripe (Chinese cane). Previously the saccharine matter was almost wholly grape sugar.

---

* Patent Office Report, 1849–50, p. 168.
† Benjamin, De Bow's Resources of S. and W. States.

9. CONDITION OF THE JUICE IN DIFFERENT PARTS OF THE STALK. *S. cane.*—In different portions of the same stalk the constituents of the juice vary considerably (at least in the case of cane grown in Louisiana), and especially in the nature of the saccharine products. In Otaheitan cane of nine months' growth, McCulloh found the three or four top joints had a sour, astringent taste, promptly staining the knife black with which they were cut, and yielding but a small quantity of juice, which, by the copper test, was found to contain grape sugar, and no starch. Subacetate of lead gave a precipitate of a dark olive color, equal in volume to $\frac{4}{5}$ of the entire mixture.

Three or four joints below these had a slightly sweet astringent and acid taste,* of a density of 7° Beaumé; with this juice subacetate of lead gave a dense, dark olive precipitate, in bulk equal to $\frac{4}{5}$ of the entire volume of the mixture as in the former case—grape sugar was also found

---

* The presence of a free acid in the juice of sorghum has induced some to infer, without investigation, that the plant contains only grape sugar. The fact is just the reverse, as is clearly proved; but that sorghum juice is not an isolated example of the fact that in the living plant the presence of an acid is not incompatible with the existence of cane sugar, the above experiment attests. The experiments of MM. Berthelot and Buignet (Comptes Rendus, vol. li. p. 1094), on the ripening of oranges, show this still more clearly. They have proved, in a series of accurate and laborious investigations, that cane sugar is formed and increases in an acid medium. Not only does the citric acid appear not to act in inverting the cane sugar already formed, but it does not oppose the augmentation of the sugar itself. The orange, both before and after the period of maturity, contains both cane and inverted sugar. The weight of the latter varies little; it was first preponderant, *but during maturation the relations changed*, and cane sugar became the most abundant. The weight of the cane sugar augments relatively to the total weight of the orange. It increases equally, whether compared with the total weight of the juice or with the weight of the fixed matter contained in the juice.—London Chem. News, 1861, vol. v. p. 117.

to be present. This juice contained also 10.4 per cent. of cane sugar.

Three or four joints from the bottom of the same cane yielded a juice of the specific gravity 10.5° Beaumé, perfectly sweet in taste, and gave no trace of grape sugar. Subacetate of lead gave a light olive-colored precipitate, in quantity much less than the similar precipitates of the above experiments, less dense, and in bulk only half the volume of the mixture. This juice contained 19.4 per cent. cane sugar.

In accordance with these facts, it is the custom in Louisiana to discard as useless the upper part of the stalk, the lower half only being sent to the mill to be pressed.

Sorghum.—In the juice of different parts of the same stalk the same constituents are found, but they bear a different ratio to each other in the different parts, and when the plant is mature, nearly all the saccharine matter which it contains throughout is cane sugar. The juice of unripe cane resembles very closely that of the upper joints of Otaheitan cane above mentioned. If a stalk of fresh ripe Chinese cane is divided transversely into three parts each of equal length, the upper joints are found to contain not only less juice, but relatively a much smaller proportion of saccharine matter than the middle or lower portions. Subacetate of lead gives a very dense and voluminous precipitate, indicating a large proportion of extractive matter. Taste unpleasant, somewhat mawkish and bitter. Juice not astringent, and does not blacken, readily, polished iron.

The middle portion of the stalk contains a less amount of impurities than the above, and its juice is more transparent. The pith is crisp and solid, possessing a pure sweet taste. The juice contains about $1\frac{1}{2}$ per cent. of matter precipitated by subacetate of lead, color of expressed juice bright olive green, perfectly colorless after precipitation by the subacetate.

The lower joints (always excepting the one which grew immediately above the surface of the ground) differ little, if at all, from the middle joints. The specific gravity of the crude juice from all parts of the stem is about the same, averaging 1·073, or about 10° Beaumé.

No portion of the stalk of sorghum cane should be rejected except the two or three upper joints, which contain but little juice.

10. CHEMICAL COMPOSITION—THE STALK. *S. cane.*—Avequin analyzed the fresh Tahiti and ribbon cane of Louisiana; Depuy, the fresh cane at Guadaloupe; Peligot, by combining the composition of cane juice with that of the dried canes sent him from Martinique, has also deduced the composition of fresh cane. The results of these analyses are compared with those of the stalk of sorghum (Chinese cane), made by Prof. J. Lawrence Smith and myself.

SOUTHERN CANE.

	Dupuy.	Peligot.	Avequin (Louisiana).	
			Tahiti Cane.	Ribbon Cane.
Sugar	17·8	18·0	14·280	13·392
Woody fiber	9·8	9·9	8·867	9·071
Mucilaginous, resinous, and albuminous matter and salts,	0·4	—	0·773	0·809
Water	72·0	72·1	76·080	76·729
	100·0	100·0	100·000	100·000

SORGHUM.

	Smith.	Stewart.
Sugar	12·0	14·5
Woody fiber, albumen, and salts	12·4	12·5
Water	75·6	73·0
	100·0	100·0

On account of the softer texture of the outer covering of the stalk of sorghum, and especially on account of the greater length of the internodes — or spaces between the joints of this cane, as compared with the tropical cane— its juice may be more readily and perfectly expressed. (See Ch. XIX.)

11. CHEMICAL COMPOSITION—THE JUICE. *S. cane.*— Properties of the juice: color, pale yellowish gray, faint balsamic odor, frothy as it flows from the mill, turbid and opalescent, on account of the suspension of finely divided matter separable by filtration. A few drops of sulphate of copper solution and an excess of caustic potash occasion, on heating, a very abundant red precipitate of suboxide of copper in the juice, indicating the presence of glucose or grape sugar (Fownes). Specific gravity 1·070 to 1·090, (Fownes, Evans), but sometimes varies from 1·046 to 1·110, or from 7° to 15° Beaumé (Evans). Otaheitan cane juice 1·070 at 77° F. (McCulloh). The juice gives a slightly acid reaction with litmus paper.

Sorghum.—Color of the fresh juice, bright green, with a tinge of yellow (Chinese cane), or yellowish green, inclining to brown in some samples (imphee)—a faint and peculiar odor most marked in imphee juice. The presence of grape sugar is generally indicated by the application of the copper test. Blue litmus paper is instantly reddened when dipped into the juice, showing that the solution is acidulous— least so in pure Chinese cane juice, most so in red imphee (Shlagoova). Specific gravity varies from 1·050 to 1·090, or from 7° to 12° Beaumé. Average of crude juice, Chinese, ripe, 1·075 to 1·080, or 10° to 11° Beaumé.

TROPICAL SUGAR CANE.

	Avequin. (La.)	Casaseca. (Cuba.)	Evans.
Sugar	15·784	20·94	18·20
Organic matter and salts	0·376	0·26	0.80
Water	83·840	78·80	81·00
	100·000	100·00	100·00

SORGHUM.

Juice of Chinese Cane. (Stewart.)	Per cent. of Sugar.	Variety of Cane.	Authorities.
Sugar ....16·00	13·7	Chinese	(Smith.)
Organic matter and salts... 1·60	16·6	Chinese	(Jackson.)
Water................82·40	15·9	Imphee	"
100·00	16	Chinese	(Madinier.)

18*

CHAPTER XXIX.

NATURAL AFFINITIES OF THESE CANES—INFERENCES DERIVED THEREFROM.

The Nature of the Relationship between Chinese Cane, Imphee, etc.
—Species—Varieties—Are they all of one Species or of several?
—Facts which throw Light upon this Question—Tests of Specific
Identity—The Inference that they are of one Species not at vari-
ance with the Fact of their wide Geographical Distribution—The
Climate of China—Early Planting of Cane necessary in that
Country—The Climate of Southeastern Africa—Inferences—
Precautions to be observed—Mr. Wray's Account of the Imphee
Canes of South Africa.

A KNOWLEDGE of the true nature of these new sugar
plants, and of their relationship to each other, would enable us to determine how far the very variable characters,
exhibited by each, are capable of being retained or fixed,
what probable limits there are to their future improvement,
and what precautions must be observed in order that the
saccharine matter in the juice may reach its most perfect
development.

The affinities of plants are determined by certain common
points of resemblance, and naturalists have generally agreed
in the opinion that, certain peculiarities, which are continually reproduced by the individual forms possessing them,
are characteristic of *Species*. In other words, species are
represented by individuals descended from a common stock,
which invariably *retain* certain peculiarities of form, etc.,
through all successive generations wherever they are found.

Very striking, but *less constant*, peculiarities of a subordinate kind are also exhibited.   Some of these, however, as long as they are subjected to human control or to highly favorable influences, are likewise regularly reproduced, and individuals, which possess them in common, are thrown into sub-groups, which are known as *Varieties*.   When the controlling influence is withdrawn, these varieties or races generally revert by degrees to their original form, losing totally, in time, their subordinate characteristics, and at last are distinguishable in no respect from their primitive type.

Thus the essential characters of species are permanent; those of varieties are inconstant.   It has also been observed, and is generally conceded, especially by botanists, that varieties freely hybridize with each other, and thus become modified, while species naturally do not; or if hybrids have in some instances been produced from two individuals of closely allied species, their offspring is barren.   The tendency of different varieties, of a single species, to freely intermix and produce an offspring as fertile as the parents is sufficiently constant, if it may not be regarded as an evidence of specific identity among all the individuals possessing it.

We are less concerned to know what is the true place of these canes in botanical classification than to determine their real affinities, and how certain peculiarities, which they now exhibit, may be perpetuated; yet, as the one involves the other, the answer must be to the question: are all the different kinds now grown in this country varieties of one species, or are they not?

If they are varieties of a single species, we may infer that they will be constantly liable to change, until their qualities have become fixed and perpetuated, through influences continually exerted upon them under human direction, or, on the other hand, that they will continually degenerate if those controlling influences are withheld.

Now what is the evidence of the facts which regularly come under our observation? We find that ever since sorghum and imphee have been cultivated within the United States, they have not only freely commingled when grown in each other's vicinity, but they have also freely amalgamated with doura, or Guinea corn, and with broom corn, and new and fertile varieties may thus readily be produced and perpetuated. No other grasses exhibit an example of such prolificacy in the hybrids of well-determined species.

But if they are distinct species, there must be found some characters to fasten upon which can be regarded as constant, and such also as demand recognition, as belonging to species. Such a character is certainly not presented in the saccharine quality of the juice, for the same is possessed in a greater or less degree by other kindred species of grasses; nor can any others properly be so regarded.

Among the most decisive tests of specific identity are the following: two races may be regarded as specifically distinct when there are no *intermediate gradations* tending to connect them, and, conversely, two races may be regarded as *specifically identical when they are connected together by close intermediate gradations.* (Carpenter.)

Applying this principle, we discover, in collections embracing the different kinds of this cane grown in the United States, that all the prominent characters, which might be thought at first sight to indicate specific diversity, are manifested in a greater or less degree by all alike, and that there is almost every possible gradation between extreme forms. Chinese sorghum, the different kinds of imphee, doura corn, and broom corn, are distinguishable from each other chiefly by such trivial characters as the

open or more compact form of the panicle, the color
and degree of prominence of the seeds beyond the glumes,
the shape of the seeds, whether round, or more or less
elliptical or oval, pubescence, difference of stature, and in
the proportions of saccharine and other matters in the juice.
None of these characters are of such a kind or importance
as to entitle them to be regarded as a basis of specific distinction.　We conclude, therefore, that like the different
varieties of Indian-corn, of the potato, the cabbage, and
the tomato, these races are all referable to a common stock.

This inference is not at variance with the fact that different varieties are now found apparently indigenous to
geographical regions very widely separated.　On the supposition that they had a common origin, there is nothing
improbable in believing that through Egypt the original
plant may readily have found its way to the extremity of
either of the connected continents, where they now grow.
But community of origin is not essential to specific identity,
and the question whether these plants originated in a single
locality or not, is unimportant in this connection.　The
wide geographical distribution of the several varieties is
more easily accounted for upon the supposition that they,
as well as other plants, were created in the regions where
they were found.

The reason why one strongly-marked variety is found
only in a remote district of Eastern Asia (Chinese cane),
and another on the southeastern coast of Africa (Imphee),
and another in Nubia and Egypt (Doura), is certainly no
more difficult of explanation than why the red currant
should be truly indigenous to both this country and Europe,
or why the wild hop-vine of our glades, or the "Jerusalem
artichoke" from the great western plains should here encounter their own living fac-similes migrating with man
from an eastern clime.

Nor is there any greater difference between the Chinese cane and the Imphee, or between the Imphee and the Doura, than the peculiarities of climate and soil are uniformly found to enstamp upon plants or animals exposed to them — or than the diversity of physiognomy of the races of men native to those regions—the Mongul and the Caffre, or the Caffre and the Copt.

The history of sorghum sugar cane is buried in such deep obscurity that it is almost useless to inquire what was the original type of the species. One variety—the Doura or Guinea corn — has been known to Europeans, perhaps, longer than any other. It was introduced into Jamaica, and thence into our Southern States, in the last century.* The brief sketch given in Chap. I. embraces nearly all the information respecting the Chinese sorghum accessible at present. The Chinese assert that they have made sugar for 3000 years. In the neighboring Islands of Japan much of the common sugar in use is made from sorghum, as we are told by the Rev. Dr. Betelheim, recently a missionary in that country. To what extent, if at all, the plant is used for the same purpose in the north of China, which has a similar climate, cannot yet be determined. From the best information hitherto obtained, the climate of Niphon (one of the Japanese Islands) and of the northern parts of China corresponds closely in temperature to the midland regions of this country in the latitude of St. Louis, Mo., or the climatic belt in which sorghum culture has heretofore proved the most successful.

In Eastern Asia, however, the geographical range of the plant northward is abruptly checked by the great rainless plateau, or Desert of Shamo, the southern limit of which follows closely the northern boundary of China, or the

---

* Morse, Geog., p. 764.

Great Wall. The whole of Northern China is drained by the river Hoangho and its tributaries, and possesses very diversified features,—lofty mountains, warm, sunny slopes, elevated plains, and broad, rich valleys. At Shanghae, which is on the coast on the southeastern border of this district, spring opens early, and the rains are periodical or at the changes of the monsoons. These occur from the middle of March to the middle of April, and from the middle of September to the middle of October. Crops are planted there before the middle of March, so as to enjoy the benefit of the spring rains. Succeeding them, or from the end of April to the middle of September, the weather is generally dry, the sky clear, hot at noonday, often rising to 100° F. in the shade during the months of July and August, and pleasant at night. These peculiarities of climate would make it necessary to plant the cane before the spring rains, or before the middle of March, and to cut it before the middle of September in this its native climate. Evidence is thus afforded that the natural habits of the plant are adapted, during the early part of its growth, to just such climatic conditions as are usual during early spring in our own country—a period of low temperature and frequent rains, succeeded rapidly by hot summer weather.

The soil* of this portion of China is said to be chiefly clay on the uplands, of a yellowish-brown color, derived from clay-slate, largely commingled with loam from the underlying rocks. The valleys are often broad and very fertile.

The correspondence of the climate of China, as well as of all Eastern Asia to that of the eastern coast of North America, is a noteworthy fact, which is due, in a great

---

* Patent Office Rep., 1857, p. 169, etc.

measure, to the physical features of the country, as well as
to the circumstance that, off the coast the warm Japan
current from the Indian Ocean, corresponding to the Gulf
Stream, tempers the climate for a great distance inland.
The flora of China and Japan is also observed to be analogous to that of the United States.*

The Zulu country of South Africa, the home of the imphees, is of a more tropical character than that of Northern
China, but is greatly modified by the variable nature of the
surface, which is low and flat along the coast, abounding
in forests and jungles, but becoming undulatory and more
elevated toward the base of the mountains which form the
edge of the great central plateau beyond, much as in the
north of China.   The flat land enjoys an almost tropical
temperature — figs, oranges, pomegranates, almonds and
cotton flourishing luxuriantly. Along the hillsides, toward
the interior, maize, melons, pumpkins, and a species of
millet, called Caffre corn, grow to a fine size. The great
diversity of climate, afforded by differences of elevation,
renders it difficult, in the absence of definite information, to
determine the natural habitat of the imphee.

Almost every variety of soil and surface is there presented, from the low-lying coast-land and the rolling
country beyond, varying gradually as the traveler ascends
the declivities of mountains clad in the foliage of temperate
climes, to the snow-line.

The mean annual temperature of this portion of Caffraria
is the same as that of the northern half of China.

One of the most important facts which have been disclosed
by the cultivation of sorghum cane in our own country, is its
great capacity for variation; some of the varieties have
markedly changed in some of their most important quali-

---

* Agassiz and Gould, Comp. Zool., p. 203.

ties,—some have deteriorated, while others have certainly
improved. It is also evident that the planter is, to a high
degree, responsible for these changes. Within certain well-
defined limits, he holds the destiny of this plant under his
control. The care and enthusiasm of the stock-raiser is
needed in this pursuit. Certain qualities may be made
transmissible, and permanent varieties, like new breeds of
cattle, may thus be originated and perpetuated. Without
sedulous attention, and enlightened skill bestowed both in
the culture of the plant and in selecting for it those conditions of growth and of improvement which its proper nature
demands, gradual deterioration will be the result; but if
the aim is to improve to the utmost every good quality,
under the guidance of a just judgment of its wants and its
capacity for improvement, we shall have varieties much
superior to any at present known.

For the purpose of comparison, Mr. Leonard Wray's
account of the different varieties of imphee cane, at the
time of their introduction into this country (the year 1857),
is here appended.

"I am acquainted with fifteen varieties of the Holcus
Saccharatus, although I doubt not there are yet others in
different parts of the world that have not come under my
notice. I shall therefore confine my remarks to the fifteen
varieties; and to prevent the constant repetition of their
botanical name I shall use their Zulu-Kaffir name of imphee
alone.

"Among Europeans, residing in South Africa, no distinction is known in regard to the varieties, and there they
will be much surprised at learning that there are really
fifteen different kinds of imphee growing before their eyes,
and being constantly eaten by them. There is certainly
that degree of similarity between them, when seen growing

19

together, which is quite sufficient to puzzle any one who
has not thoroughly studied them, and this is so much the
case that there are very few male Kaffirs even who can dis-
criminate between some of the varieties; in consequence
of which I had very frequently to call in the superior agricultural knowledge of the old Kaffir women, until I had
learned to distinguish them with certainty myself.

"When I had attained this first most desirable step, I
had next to learn their several peculiarities and value,
which I was able to do with greater exactness by planting
the seed, watching them daily during their growth, and
eventually testing the saccharine value of their juices, and
making sugar from them all in large quantities.

"*Vim-bis-chu-a-pa.* This is the largest size, and the
tallest of the whole, while it is full of juice and very sweet.
When planted in rich alluvial soil it attains its greatest
size and most perfect development, requiring from four to
five months to arrive at maturity.   It grows to a height of
from ten to fifteen feet, is one and a half to two inches in
diameter at the lower end of the stalk, and usually cracks
or splits as its ripens.   By means of a most primitive and ill
constructed little wooden mill, I obtained sixty per cent. of
juice from the stalk.   This juice was clean and clear, and
the saccharometer showed it to contain fourteen per cent.
of sugar, after I had removed the feculæ by means of cold
defecation.   The sugar it yielded was fully equal to the
best cane sugar of the West Indies.   The stalks, carefully
weighed, were found to vary from one and a half to two
and a half pounds English weight each, trimmed ready for
the mill.   The seed head, which is very large and beautiful,
is generally from twelve to eighteen inches in length, containing many thousands of fine plump seeds, of a sandy,
yellow color, strongly held by a sheath, which partially envelops them.

" *E-a-na-moo-dee* is the next in size, and is very similar both in habit and value to the last. It attains a height of twelve to thirteen feet, but is not so coarse in appearance, nor does it contain so much woody fiber as the Vim-bis-chu-a-pa, but is rather softer and more juicy, I having obtained from it sixty-four per cent. of juice, containing fourteen per cent. of sugar. The stalks weigh from one to two pounds when trimmed ready for the mill, and I have cut as many as eleven such stalks from one root or stool. The seed heads are large, but stiff and erect, containing quantities of large, round, plump seeds, of a clear yellow color. In general they may be said to ripen two weeks earlier than the last named. Like the Vim-bis-chu-a-pa, this variety ratoons in about three to three and a half months after the first cutting.

" *E-en-gha.* This is a fine, tall kind, being from ten to twelve feet high when full grown, but it is more slender than either of the foregoing, and exceedingly graceful in appearance. It begins flowering in ninety days, and is fully ripe three weeks after; we will, therefore, class it at four months. I have had stalks weighing as much as one pound fourteen ounces each. The largest commonly obtained may then be estimated at two pounds weight; yielding, by my poor little mill, sixty-eight per cent. of juice, containing fourteen per cent. of sugar. I have obtained ten stalks from one stool. They ratoon in three months after cutting. The seed head of the E-en-gha is large and very pretty, the seed being upon long, slender foot-stalks, which are bent down by the weight of the seed, forming a graceful drooping. The seeds, which are of a dull yellow color, are rather long and flat than plump and round.

" *Nee-a-za-na* " (White Imphee) " is held by the Zulu-Kaffirs to be the sweetest of all the imphee kind; but I

found the Boom-vwa-na and the Oom-see-a-na quite as
sweet, and, in my estimation, their juices are superior to it
in some points. My Zulus have told me that under favorable circumstances the Nee-a-za-na frequently ripens in
seventy-five days; and my head-man (a most intelligent
native plowman) declares that he has had them from his
own land as sweet as any sugar cane. From my own
actual experience I found that they ripened in about three
months, and that they were the softest and most abounding,
in juice of any. With my mill, I obtained seventy per
cent. of juice, much still remaining in the trash, and the
saccharometer showed fifteen per cent. of sugar after
cold defecation. This, then, for European culture is a
perfect gem of a plant; one which will be anxiously
sought after, and very generally cultivated throughout
Europe at least. In two months after the first cuttings I
have had the ratoons six feet high, and in flower. The Nee-
a-za-na is a very small sized variety, but tillers out greatly,
having sometimes fifteen stalks to one root. I have had
its stalks varying from four ounces to twelve ounces in
weight, but they *do* grow rather larger than this. It
always appeared to me that their juice was more mucilaginous, and abounding in faculæ than the two varieties I
have just mentioned. The seed head is very bushy and
branchy, and, when thoroughly ripe, the seeds are large,
round and plump.

" *Boom-vwa-na* is a most excellent and valuable variety, .
of which I have eaten single pieces, containing certainly
two or three per cent. more sugar than the average juice
obtained from large bundles of stalks taken as they come.
This average juice never contained less than fifteen per
cent. of sugar, as indicated by the saccharometer, after
the raw juice had been cold defecated; and there is a clear-
ness, a brightness, and a genuine sugar cane sweetness in

the juice of this variety, and of the *Oom-see-a-na* that I very much admire. In its growth and general appearance it is very much like the E-en-gha, but its stalks are brighter and more slender; its leaves are not so broad, and its seed vessels are upon shorter and stiffer foot-stalks. The stalks have a pinkish-red tint, which increases as they approach maturity; and the seed cases have a pink and purple hue, mixed with the general yellow ground. The Boom-vwa-na tillers very much, giving from ten to twenty stalks for one root; but they seldom weigh more than one pound each. I have obtained seventy per cent. of juice, which is easily clarified, and makes a beautiful sugar. The plant reaches perfection in from three to three and a half months.

"Oom-see-a-na" (a variety of this misnamed "Ota-heitan") "is a peculiarly marked variety, in consequence of the purple or black appearance of its seed heads, arising from the sheath or seed cases being of this color, and not the seed itself. The seed head is very stiff and erect, with short strong foot-stalks, and the seed is large, round and full. In time of growth and goodness of juice, it is very similar to the Boom-vwa-na; its stalks are small, but numerous. They both ratoon well from first cuttings.

"*Shla-goo-va*" (Red Imphee) "is slightly inferior to the three last named varieties, but is nevertheless very valuable, and much prized by the Zulus. It takes three and a half months to ripen, and becomes a tall, good sized plant, but its chief distinction is the exceeding beauty and elegance of its seed heads. The foot-stalks are extremely long, which causes them to have a graceful drooping, while the seed cases, or sheaths, vary in color from a delicate pink to a red, and from a light to a very dark purple, but each color very bright and glistening, forming on the whole an extremely beautiful appearance.

"*Shla-goon-dee*. This is a sweet and good variety, and
19*

under favorable conditions produces fine-sized stalks. The
seed heads are very stiff and erect, and the seed vessels
are compact and very close. It usually takes three and a
half months to reach maturity, and it ratoons very quickly,
as the following memorandum of my diary will show:
'December 13th. Cut down a small patch of imphee, and
dug up the ground for the purpose of planting arrow-root,
but some of the imphee not being entirely eradicated,
sprung up afresh, some roots having fifteen stalks each.
On the 18th of February, one of them (Shla-goon-dee)
was upwards of six feet in height, with a thick stalk, and
the seed head just thrown out, *being only two months and
five days* from its being cut down and apparently destroyed.' This bunch of seed I gathered during the first
week in March, and I have it now in England.

"*Zim-moo-ma-na.* This is likewise a sweet and good
variety, with seed heads upright and compact, and fine
plump seeds, very numerous.

"*E-both-la, Boo-e-a-na, Koom-ba-na, See-en-gla, Zim-
ba-za-na,* and *E-thlo-sa,* form the remainder of the fifteen
varieties, each differing slightly from the others in its
saccharine qualities, as well as in appearance, but still
easily distinguished from each other by any one who has
studied them."

It will be seen from the above descriptions, that the
imphees now most commonly grown in this country have
already suffered considerable modification. The Neeazana and Oomseeana have improved.

CHAPTER XXX.

OTHER PRODUCTS.—SYNOPSIS OF PROCESS OF MANUFACTURE.

Success in Sugar Manufacture in a great degree dependent upon
Economy in the Use of the Waste Products—Value of the Cane
Trash for Manure—Paper Manufacture—Roofing—How Silk,
Wool, and Cotton may be dyed various Colors from a Material
contained in the Seeds and Trash—Use of defecated unripe
Juice as a Source of Grape Sugar, to add to Grape Wine by Dr.
Gall's Method—Uses of the Seed, etc.—Vinegar.
Synopsis of the Process of Manufacture—Cream of Lime—Mode of
Purifying Barrels.

THE utilization of what are commonly regarded as the
waste products of a sugar manufactory is of much importance. Although the capacity of the plant for the production of sugar and syrup is to be regarded as the principal
and almost exclusive element in any rational computation
of its value, we should not lose sight of the fact that success
in this branch of industry at the North will depend in a
great degree upon the skill of the sugar planter, in making
the best possible use of every material within his reach.

It is of primary importance to him to increase the productiveness of his lands to the utmost extent within his
power. It has already been shown (Ch. VII.) how the
trash pile may readily be made subservient to that end by
restoring annually to the land most of the substances necessary to the growth of future crops, which are annually

(223)

removed with the cane. This use of the trash and of all other waste materials, derived from the plant or introduced during the process of manufacture, will generally be found preferable to any other uses to which they may be applied. It has been found that a cheap and good quality of paper can be produced from the crushed cane, and where a paper mill is convenient, it may be more economical to dispose of it for that purpose, provided that a larger quantity of materials, as valuable for manure as the trash, is purchased and returned again for the price for which the latter was sold.

There is another use, however, to which cane trash may be put, that well deserves attention. It has been known for some years past, that the Chinese produce in fabrics of silk and wool a beautiful red color, which is derived from the seeds of sorghum. Experiments made both in this country and in Europe, likewise show that the same dye may be obtained from the crushed canes. For this purpose they must be sheltered from the rains after they are received from the mill, and thrown in close piles until fermentation sets in. Afterward the heaps are to be opened, and stirred frequently to prevent heating, which would destroy the coloring matter. When their color has changed to red or reddish brown, they are then cut up, washed and dried. A weak lye of caustic potash may be used to extract the color from them.* By neutralizing this alkaline solution with a weak solution of oil of vitriol the color falls in the form of red flakes, which are easily soluble in alcohol, alkalies, and diluted acids.

To Mr. Henri Erni, Chemist in the National Agricultural Department, we are indebted for the following obser-

---

* A. Winters, in Liebig's Jahresbericht, 1859, p. 754; quoted by Wetherill, Ag. Rep. 1862, p. 535.

vations respecting the mode of using this new dyeing material, derived from experiments conducted by himself in the laboratory at Washington City:

* "The simplest solvent is alcohol (very expensive at present). Dilute acids were resorted to with very good success, and at an expense and trouble hardly worth mentioning. The seeds were boiled in vinegar, or in water to which oil of vitriol had been added, before heating, until the mixture tasted as acid as vinegar. Other acids, such as tartaric, oxalic, etc., can be used, but are more expensive. When the liquid assumes a red or rather an intense orange color it is ready for use. The articles to be colored are at once brought into the hot solution, and agitated until the color no longer increases. They are at once removed and dipped into a weak solution of salt of tin (chloride of tin), obtained by dissolving tin in hydrochloric acid. They are then exposed to the air for a short time, and washed.

"Cotton and silk may thus be colored red, wool turns to a beautiful purple, and an almost unlimited variety of colors and shades may be obtained by substituting for salts of tin other mordants. All the various shades of red, purple, orange, gray, etc. are thus produced from the same bath, the cloths being afterward drawn through solutions of protochloride of tin, bichromate of potash, sulphate of copper, ammonia, lime-water, subnitrate of bismuth, etc. Yellow is produced by adding to the seed sufficient nitric acid to form a thick mushy mass. Too much acid will make a straw color.

"The dye turns solid by standing, and may thus be stored. To dye silk or wool yellow, the solid dye is dissolved in boiling water, the goods dipped into it and afterward washed.

---

* Report of Department of Agriculture, 1864, p. 532.

" Cotton has the least attraction for sorghum dyes, while wool receives the brightest colors. The same dye is developed in the stalks."

The employment of trash for thatching has already been alluded to. (Ch. XIII.)

The addition of the juice of unripe sorghum, when it is thoroughly defecated and made more dense by evaporation, to grape-must naturally deficient in sugar, so that wine of a good quality can be made of it according to the method of Dr. Gall, has been suggested, and is already practiced in some parts of France.

The grape, in seasons unfavorable to its ripening, or when grown in a climate where its best qualities cannot be developed, produces a wine which contains too little of sugar, and consequently of alcohol, and which will not keep, but runs directly into the acetous fermentation. But when a sufficient amount of grape sugar, or sugar of the same kind as that found naturally in the ripe grape, is added to the must, it becomes converted into a wine of as good a quality, when properly made, as that which is produced in a more genial climate or favorable season.

It is just this species of sugar which the juice of unripe cane contains,—and with this source at hand from which it can be derived, it places the wine-grower in some degree independent of the vicissitudes of the seasons, and makes it within his power to produce a wine in an uncongenial season of nearly as good a quality as in favorable years. He has merely to add to the juice of the grape a certain proportion of grape sugar in solution, in order that the fermentation may not be completed until a sufficient amount of alcohol has been produced to bring the wine up to the proper standard of strength and quality. To be used for this purpose, the juice must be thoroughly defecated, so as to free it from gum and all injurious ingredients. Wines thus prepared differ but little from those prepared from well-ripened grapes, and are much to be preferred to the fermented fruit juices in common use.

Sorghum seed has frequently been used for the manufacture of a kind of flour which has proved to be nutritious and valuable. It somewhat resembles buckwheat flour. In the Southern States particularly, large quantities of it were used during the late rebellion. That made from certain varieties of cane contracts a light pinkish color from the outer covering of the seed. The seed, ground with oats or corn, or without intermixture, forms a valuable feed for horses and cattle, and is considered to be equal to rye for this purpose. The whole grain boiled is also excellent food for cattle or fowls. From thirty to fifty bushels of seed may be grown upon an acre of ground. The blades, and young cane from the second cutting of early ripening varieties, may profitably be cured (see Ch. XII.) and used as hay.

The scum produced in this process is very valuable for feeding swine for fattening, but the precaution must be taken to commence feeding it in small quantities, gradually increased, and always in connection with grain. It should not be fed in a fermented condition.

Ordinary cane juice is capable of being converted into from 5 to 9 per cent. of alcohol. The mode of making vinegar from sorghum juice, or from the washings of the pans, the filter, etc., differs in no respect from that ordinarily pursued in converting cider into vinegar, except that when the juice of cane is used, it must previously have been thoroughly defecated. Before fermentation a cupful of fresh yeast must be added to each cask.

Into one division of the double tank M, admit the juice from the mill through the pipe (Fig. 5) represented below,

Fig. 5.

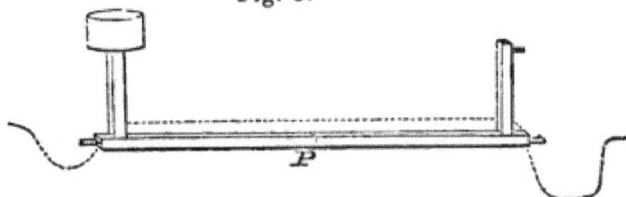

and placed as indicated in Ch. XV., adopting the precautions there mentioned.* If the mill is placed upon a platform (Ch. XXV.), a straight pipe only is required

Fill up one division of the tank to the mark, and turn the current of juice into the other. Add gradually to the former milk of lime until the red litmus paper changes color, stirring the juice well. Then pour in the proper measure of the clarifier, and stir. Meanwhile, the fire has been kindled in the furnace. When the water in the evaporator is boiling briskly, open the gate of division A for its escape into B. At the same time permit the prepared

---

* For the bucket upon the head of the pipe a larger vessel may be substituted (if preferred), consisting of a tightly jointed box about 2 feet in length by 15 inches in width and the same in depth. Two or three light frames, covered with coarse canvas, fitting into grooves in the sides of the box, divide it transversely into compartments, through which the juice passes successively, entering the box at one end and passing out into the pipe at the other. All the suspended matters in the liquid are thus removed, and with the advantage that the canvas strainers in the upright position are not easily clogged, and are removed with facility when it is necessary to clean them. The first of the series of strainers should be coarse, so as to arrest the grosser impurities, such as fragments of pith, etc., and the rest successively finer. A modification of this simple arrangement constitutes the celebrated *leaf filter* of Mr. Jos. S. Lovering.

juice from the tank to enter A in a small continuous stream. (See directions, Ch. XV.) Draw off the water successively from all the divisions of the pan, and admit in its place the clarified juice. The boneblack filter should previously have been carefully filled with well washed, odorless (well burned) animal charcoal. Retain the water upon the filter, so that it may cover its surface, until the syrup is to be admitted into C, when it must be allowed to escape in advance of the juice. Admit the syrup into the finishing pan D, at intervals, in sufficient quantities, but not to a greater depth than $1\frac{1}{2}$ inch at a time. Boil for syrup to 228° F., or for sugar to 230°–232° F.

For the defecation of juice which is more than ordinarily impure or unripe, the first method indicated in Ch. XX. is perhaps to be preferred, but the second method should generally be employed when the cane is in good condition.

Remember to *keep the depth of juice in A at about $\frac{1}{2}$ inch,* and never flood it. *The central square between the ledges should be continually overspread by a sheet of boiling juice.* Do not permit the filter to be overflowed, but allow the syrup in the filter to cover constantly the surface of the charcoal.

Syrup for crystallization should be allowed to accumulate in the cooler, and should be removed at night to a room kept heated to not less than about 80° F. Observe the precautions to be used in filling the boxes, draining, etc. (Refer to Ch. XVII., etc.)

Cream of Lime. This should always be prepared beforehand, from good quick-lime, as follows : Upon a half peck of lime, freshly slacked, pour two or three gallons of boiling water, stir the mixture, and let the lime have time to settle to the bottom ; decant the clear liquid ; pour

20

upon the lime another gallon or two of hot water, and decant a second time. Place the washed lime in a deep jar, mixed with just enough of water to make a mixture of the consistence of very thick cream. The coarse particles will fall to the bottom, while the finer sediment will occupy the upper part of the jar. Dilute it with a little water when it is to be used.

Cleansing of Barrels. Syrup barrels of all kinds, and especially those which have contained old molasses, should be thoroughly cleansed. The importance of this will be better appreciated when it is known that if it be neglected or imperfectly performed, a refined syrup will acquire from the barrel a flavor which will render useless all the care and skill previously bestowed upon it. The barrels should first be washed out by pouring hot water into them, turning them frequently, and standing them up on each end alternately until the water has abstracted all that it is capable of dissolving. Lime-water may then be poured in, the bung inserted, and the casks turned at intervals, for a day or two, as before. Finally rinse them out, and suspend into each barrel, upon a bent wire, a lighted sulphur match, made beforehand by dipping a piece of cotton cloth into melted brimstone. A narrow slip of muslin may be so prepared, and afterward cut into pieces of three or four inches square, one of which is sufficient for fumigating a single cask. Insert the bung while the match is burning. Afterward withdraw the wire, and keep the barrel bunged up until just before it is to be used, when it should be carefully rinsed out.

New casks of oak may be purified simply by introducing into them a few gallons of lime-water, or a smaller quantity of thin milk of lime, and rolling them about frequently, so that the whole of the interior may be saturated with the lime solution. Wash out with a little fresh water when they are wanted for use.

A mouldy smell or *tang* may be removed by introducing a peck of wheat or rye bran, mixed with a little warm water, into the barrel, and add to it a cupful of fresh yeast. Insert the bung loosely, and let the mixture remain in the barrel for a week. Turn the barrel occasionally, and during the fermentation which ensues; the mouldiness will be entirely removed. The fermentation will be more rapid if the cask is kept in a moderately warm place.

I.

TABLE "Showing the quantity of sugar contained in one hundred pounds of expressed cane juice, or syrup, of good quality, and also of the quantity of water that must be evaporated, to reduce the same to the state of saturated syrup at each degree. A saturated solution of very pure sugar contains five parts of sugar and three parts of water. This is indicated by 34° of Beaumé's saccharometer, at the temperature of 82° Fahr."—*Dutrone.*

Degrees of density by Beaumé's Scale.	Weight of Sugar in each 100 lbs. of Juice or Syrup.			Weight of Water in each 100 lbs. of Juice or Syrup, beyond the water of solution.		
	lbs.	*oz.*	*dr.*	*lbs.*	*oz.*	*dr.*
1	1	13	6	97	...	15
2	3	10	12	94	1	14
3	5	8	3	·91	2	13
4	7	5	10	88	3	12
5	9	3	...	85	4	11
6	11	...	7	82	5	10
7	12	13	14	79	6	9
8	14	11	4	76	7	8
9	16	8	11	73	8	7
10	18	6	1	70	9	6
11	20	3	8	67	10	5
12	22	...	15	64	11	4
13	23	14	5	61	12	3
14	25	11	12	58	13	3
15	27	9	2	55	14	1
16	29	6	9	52	15	1
17	31	4	...	50	...	...
18	33	1	6	47	...	15
19	34	14	13	44	1	14
20	36	12	3	41	2	13
21	38	9	10	38	3	12
22	40	7	1	35	4	11
23	42	4	7	32	5	10
24	44	1	14	29	6	9
25	45	15	4	26	7	8
26	47	12	11	23	8	7
27	49	10	1	20	9	6
28	51	7	8	17	10	5
29	53	4	15	14	11	4
30	55	2	5	11	12	3
31	56	15	12	8	13	2
32	58	13	3	5	14	1
33	60	10	9	2	15	...
34	62	8	...	...	...	...

II.

TABLE by Dr. Evans, showing the per cent. of sugar in solutions of different degrees of density, according to the scale of Beaumé, and corresponding specific gravities.

Degrees of density, Beaumé.	Sugar in 100 parts.	Specif. grav.	Degrees of density, Beaumé.	Sugar in 100 parts.	Specif. grav.
1	·018	1·007	19	·352	1·152
2	·035	1·014	20	·370	1·161
3	·052	1·022	21	·388	1·171
4	·070	1·029	22	·406	1·180
5	·087	1·036	23	·424	1·190
6	·104	1·044	24	·443	1.199
7	·124	1·052	25	·462	1·210
8	·141	1·060	26	·481	1·221
9	·163	1·067	27	·500	1·231
10	·182	1·075	28	·521	1·242
11	·200	1·083	29	·541	1·252
12	·218	1·091	30	·560	1·261
13	·237	1·100	31	·580	1.275
14	·256	1·108	32	·601	1·286
15	·276	1·116	33	·622	1·298
16	·294	1·125	34	·644	1·309
17	·315	1·134	35	·666	1·321
18	·334	1·143			

III.

TABLE showing how a saturated solution of sugar is affected by being reduced to different degrees of density, as indicated by the thermometer, commencing at the point of saturation. This table was prepared with great care by Dutrone from actual experiments. By means of it the number of pounds of sugar, which will crystallize in syrup of any given density, as determined by its temperature at the boiling point, may be accurately predicted, and the

20*

number of pounds of water which it has lost by evaporation between the temperature of saturation and the given temperature. Also the number of pounds of sugar and water respectively, which remain combined in the form of syrup (drippings). At 81·5° F. (22° Reum.) Dutrone found that three parts of water and five parts of sugar form a saturated solution. He formed a saturated syrup therefore consisting of 60 pounds of water and 100 pounds of sugar, and upon this the table is based.

If 160 pounds of such syrup be raised from the boiling point, 219° F. to 232° F. (for example), 33 lbs. 11 oz. 10 dr. of water will have evaporated; 56 pounds of sugar will crystallize when the syrup is cooled to the proper temperature and treated in the regular manner, and 26 lbs. 4 oz. 6 dr. of water and 44 lbs of sugar still remain combined.

To determine this it is necessary only to have a thermometer which will register accurately the temperature of the boiling syrup.

Degree of therm'r. Fahrenheit.	Weight of Sugar which separates in crystals.			Weight of Water evaporated.			Weight of Sugar yet combined with water in the state of syrup.			Weight of Water still combined with sugar in the state of syrup.		
	lbs.	oz.	dr.	lbs.	oz.	dr.	lbs.	oz.	dr.	lbs.	oz.	dr.
219	0	......	......	0	......	......	100	......	......	60	......	......
221	8	......	......	4	12	14	92	......	......	55	3	2
223	19	4	......	11	8	14	80	12	......	48	7	2
225·5	30	......	......	18	......	......	70	......	......	42	......	......
228	41	......	......	24	9	10	59	......	......	35	6	6
230	52	......	......	31	3	4	48	......	......	28	12	12
232	56	......	......	33	11	10	44	......	......	26	4	6
234·5	60	5	......	36	3	......	39	11	......	23	13	......
237	63	4	......	38	1	......	36	12	......	21	15	......
239	66	3	......	39	4	......	33	13	......	20	12	......
241	69	2	......	41	7	10	30	14	......	18	8	6
243·5	72	1	......	43	4	......	27	15	......	16	12	......
246	75	......	......	45	......	......	25	......	......	15	......	......
248	77	7	......	46	7	4	22	9	......	13	8	12
250	80	5	......	48	7	8	19	11	......	11	8	8
252·5	83	3	......	50	1	10	16	13	......	9	14	6
255	85	......	......	51	......	......	15	......	......	9	......	......
257	87	4	......	52	5	14	12	12	......	7	10	2
259	88	6	......	53	1	6	11	10	......	6	14	10
261·5	90	1	......	54	1	......	9	15	......	5	15	......
264	91	4	......	55	3	10	8	12	......	4	12	6
266	92	7	......	55	12	......	7	9	......	4	4	......
268	94	2	......	56	7	10	5	14	......	3	8	6
270·5	95	5	......	57	3	8	4	11	......	2	12	8
273	97	......	......	58	6	8	3	......	......	1	9	8
275	98	2	......	58	14	8	1	14	......	1	1	8
277	99	2	......	59	7	10	14	......	......	......	8	6
279·5	100	......	......	60	......	......	......	......	......	......	......	......

IV.

PELIGOT'S TABLE *for determining the per cent. of sugars in a solution by saccharate of lime*

Quantity of sugar dissolved in 100 parts of water.	Density of Syrup.	Density of Syrup when saturated with lime.	100 parts of residue dried at 120° (Cent.) contain—	
			Lime.	Sugar.
40·0	1·122	1·179	21·0	79·0
37·5	1·116	1·175	20·8	79·2
35·0	1·110	1·166	20·5	79·5
32·5	1·103	1·159	20·3	79·7
30·0	1·096	1·148	20·1	79·9
27·5	1·089	1·139	19·9	80·1
25·0	1·082	1·128	19·8	80·2
22·5	1·075	1·116	19 3	80·7
20·0	1·068	1·104	18·8	81·2
17·5	1·060	1·092	18·7	81·3
15·0	1·052	1·080	18·5	81·5
12·5	1·044	1·067	18·3	81·7
10.0	1·036	1·053	18·1	81·9
7 5	1·027	1·040	16·9	83·1
5·0	1·018	1·026	15·3	84·7
2·5	1·009	1·014	13·8	86·2

V.

TABLES OF WEIGHTS AND MEASURES *officially recognized in the United States by recent Act of Congress.*

MEASURES OF LENGTH.

Metric Denominations and Values.		Equivalents in Denominations in use.
Myriameter ...	10,000 meters,	6·2137 miles.
Kilometer	1,000 meters,	0·62137 mile, or 3,280 feet and 10 inches.
Hectometer	100 meters,	328 feet and one inch.
Dekameter.....	10 meters,	393·7 inches.
Meter............	1 meter,	39·37 inches.
Decimeter......	$\frac{1}{10}$th of a meter,	3 937 inches.
Centimeter	$\frac{1}{100}$th of a meter,	0·3937 inch.
Millimeter......	$\frac{1}{1000}$th of a meter,	0·0394 inch.

MEASURES OF SURFACE.

Metric Denominations and Values.		Equivalents in Denominations in use.
Hectare.........	10,000 square meters,	2·471 acres.
Are	100 square meters,	119·6 square yards.
Centare.........	1 square meter,	1550 square inches.

MEASURES OF CAPACITY.

METRIC DENOMINATIONS AND VALUES.			EQUIVALENTS IN DENOMINATIONS IN USE.	
Names.	No. of liters.	Cubic Measure.	Dry Measure.	Liquid or Wine Measure.
Kiloliter or stere	1,000	1 cubic meter..........	1·308 cubic yards....	264·17 gallons.
Hectoliter...	100	$\frac{1}{10}$th of a cubic meter.............	2 bus. & 3·35 pecks..	26·417 gallons.
Dekaliter....	10	10 cubic decimeters..	9·08 quarts...........	2·6417 gallons.
Liter	1	1 cubic decimeter	0·908 quarts...........	1·0567 quarts.
Deciliter	$\frac{1}{10}$	$\frac{1}{10}$th of a cubic decimeter	6·1022 cubic inches	0·845 gill.
Centiliter....	$\frac{1}{100}$	10 cubic centimeters.............	0·6102 cubic inch....	0·338 fluid oz.
Milliliter	$\frac{1}{1000}$	1 cubic centimeter...	0·061 cubic inch......	0·27 fluid drm.

WEIGHTS.

METRIC DENOMINATIONS AND VALUES.			EQUIVALENTS IN DENOMINATIONS IN USE.
Names.	Number of grams.	Weight of what quantity of water at maximum density.	Avoirdupois Weight.
Millier or tonneau.......	1,000,000	1 cubic meter..................	2204·6 pounds.
Quintal	100,000	1 hectoliter.......................	220·46 pounds.
Myriagram.................	10,000	10 liters........................	22·046 pounds.
Kilogram or kilo........	1,000	1 liter............................	2·2046 pounds.
Hectogram.................	100	1 deciliter......................	3·5274 ounces.
Dekagram..................	10	10 cubic centimeters........	0·3527 ounce.
Gram........................	1	1 cubic centimeter...........	15·432 grains.
Decigram...................	$\frac{1}{10}$	$\frac{1}{10}$th of a cubic centimeter	1·5432 grains.
Centigram..................	$\frac{1}{100}$	10 cubic millimeters..........	0·1543 grain.
Milligram..................	$\frac{1}{1000}$	1 cubic millimeter............	0·0154 grain.

INDEX.